SURFACE AND SUBSURFACE RUNOFF GE
IN A POORLY GAUGED TROPICAL COA

A STUDY FROM NICARAG

SURFACE AND SUBSURFACE RUNOFF GENERATION PROCESSES IN A POORLY GAUGED TROPICAL COASTAL CATCHMENT

A STUDY FROM NICARAGUA

DISSERTATION

Submitted in fulfillment of the requirements of

the Board for Doctorates of Delft University of Technology

and of

the Academic Board of the UNESCO-IHE Institute for Water Education

for the Degree of DOCTOR

to be defended in public on

Friday, 16 January 2015 at 12:30 hours

in Delft, The Netherlands

by

Heyddy Loredana Calderon Palma

Master of Science in Hydrogeology, University of Calgary, Canada

Chemical Engineer, National University of Engineering, Nicaragua

born in Nicaragua

This dissertation has been approved by the promotor:

Prof. dr. S. Uhlenbrook

Composition of Doctoral Committee:

Chairman	Rector Magnificus TU Delft
Vice-Chairman	Rector UNESCO-IHE
Prof. dr. S. Uhlenbrook	UNESCO-IHE / TU Delft, promotor
Prof. dr. M.E. McClain	UNESCO-IHE / TU Delft
Prof. dr. ir. H.H.G. Savenije	TU Delft /UNESCO-IHE
Prof. dr. B. Diekkrüger	University of Bonn, Germany
Prof. dr. ir. P. van der Zaag	UNESCO-IHE / TU Delft
Dr. G.M. Gettel	UNESCO-IHE, advisor
Prof. dr. ir. A.E. Mynett	UNESCO-IHE / TU Delft, reserve member

This research was conducted under the auspices of the Graduate School for Socio-Economic and Natural Sciences of the Environment (SENSE)

Cover illustration by Serena Mitnik-Miller

CRC Press/Balkema is an imprint of the Taylor & Francis Group, an informa business

Published by:
CRC Press/Balkema
PO Box 11320, 2301 EH Leiden, The Netherlands
e-mail: Pub.NL@taylorandfrancis.com
www.crcpress.com – www.taylorandfrancis.com

ISBN 978-1-138-02758-9 (Taylor & Francis Group)

I want to know,
Have you ever seen the rain?
I want to know,
Have you ever seen the rain
Comin' down on a sunny day?

John Fogerty, 1971.

To my life mentors:

My mom, who taught me the joy of reading;

My dad, who showed me the passion for learning;

My sister, who was my first role model.

Acknowledgements

This book is the result of a long sustained effort through many and different challenges. All of which shaped my intellectual and personal development during the process of completing my PhD. My effort was not unaccompanied, of course; it relied on God's blessings, the love of my family and friends; and the support of numerous extraordinary institutions and people.

I have to thank first, the National Autonomous University of Nicaragua (UNAN-Managua), through the people who allowed me to pursue this goal. Former Rector, Prof. Francisco Guzman, actual Rector Prof. Elmer Cisneros; Luis Medina and Lorena Pacheco from the Vice-Rectory of Research and Graduate Studies.

Many thanks to the former Director of the Aquatic Resources Research Center (CIRA-UNAN), Prof. Salvador Montenegro-Guillén and former Deputy Director Dr. Katherine Vammen, who supported all my efforts and helped me to knock on the right doors. I want to thank all my colleagues at CIRA, who always welcomed with a smile my unscheduled visits to borrow equipment and materials for water sampling and gave me many practical and useful advises on this matter. Many thanks to my colleagues from the Hydrogeology Department, who cheered up my work in Nicaragua. I would like to specially thank the Head of the Department and dear friend, Yelba Flores, who came along to long walks in the catchment, providing geological expertise and always helping me to keep up the spirit.

My gratitude goes also to the Director of the Institute of Geology of Geophysics (IGG-UNAN), Dr. Dionisio Rodríguez, who supported my research with geophysical and drilling equipment, as well as technical and scientific personnel who helped me tremendously in the field. Special thanks to my colleagues and friends Marvin Corriols and Lener Sequeira who provided geophysical expertise.

Also I want to thank the Municipality of San Juan del Sur, which welcomed my research and provided valuable information to begin my work; in particular, to Mr. Bayardo Romero. Also, to the many people from Ostional who helped me with field work, who guided me throughout the catchment and shared their homes and meals with me. I am very grateful to my field assistants, whom shared my enthusiasm and sometimes my disappointments, during the ups and downs of my field work.

I am very thankful to the UNESCO-IHE staff, who provided the conditions for my work during my stays in the Netherlands, in particular to Jolanda Boots and Tonneke Morgenstond. Many thanks to the whole lab staff, who helped me and trained me in analytical procedures and gave a word of advice when needed. Special thanks also to my friend Dr. Gerald Corzo for many insightful conversations and words of advice and encouragement.

I need to thank Dr. Mary C. Hill from the USGS and Dr. Lawrence R. Bentley from the University of Calgary, who made possible my attendance to an IAHS conference in Brazil, where I had the chance to meet, among other inspiring hydrologists, Prof. dr. Stefan Uhlenbrook. My special gratitude goes to him, who gave me the opportunity to work under his supervision, helped me channel my research through a tortuous path, always encouraging my work and providing decisive feedback to help me find my own pathway.

Thanks also to all the friends and colleagues whom I encountered along these years at UNESCO-IHE. Most of them already scattered around the world but who left an enduring impression in my academic and personal life. Special thanks to my dear friend Gabriela Alvarez for always helping me see the brighter side of life.

All these efforts relied on the generous financial support from my sponsors. I am very grateful to The Netherlands Fellowship Program (Nuffic). I would also like to thank the International Foundation for Science (IFS). Finally, my deepest gratitude for the generous and decisive financial support from the Faculty for the Future program (www.facultyforthefuture.net); and the encouragement and support provided by their staff and my fellow grantees.

Heyddy Calderon
Delft, The Netherlands

Summary

Hydrological research in the humid tropics is particularly challenging because of highly variable hydrological conditions. These regions are also under high socio-economic stresses caused by rapid population increase, which leads to land use changes. Additionally, climate change and variability also induce changes in the hydrologic regime. Nevertheless, understanding of hydrological processes in these areas is limited and transfer of hydrological knowledge from other hydro-climatic regions to humid tropical catchments may be difficult due to their intrinsic differences. This is especially problematic for developing countries, where limitations to produce reliable predictions impair sustainable management of water resources. Central America, and Nicaragua in particular, are good examples of these regions.

The objective of this research is to understand the surface and subsurface runoff generation processes in a poorly gauged coastal catchment in Nicaragua under humid tropical conditions. Specifically, this research focuses on identifying geomorphological and hydro-climatic controls on catchment response at different spatio-temporal scales; studies the link between hydrological processes and ecosystem conditions (*i.e.* mangrove forest); and analyzes the significance of runoff generation processes for water resources management.

The study area shares the topographic, geologic and hydro-climatic characteristics of other catchments on the South Pacific of Nicaragua. Land use commonly includes forests, agriculture and cattle grazing. Mangrove ecosystems are typically found in these catchments. Population mostly relies on shallow groundwater for water supply, and sanitation systems are missing. However, the South Pacific Coast of Nicaragua has great touristic potential and real estate development is occurring quickly. The increase in tourism and other related developments will further increment the stress on water resources in this region.

This thesis is organized as follows: Chapter 1 provides an overview of water resources management in poorly gauged catchments and hydrological challenges in the humid tropics. It also outlines the water resources situation in Nicaragua, and it describes the study area. Chapter 2 investigates groundwater flow systems using a combination of geophysical, hydrochemical and isotopic methods. Electrical resistivity tomography (ERT) was applied along a 4.4 km transect parallel to the main river channel and in five cross sections, to identify fractures and determine aquifer geometry. Stable water isotopes, chloride and silica were analyzed for springs, river, wells and piezometers samples during the dry and wet season of 2012. Indication of moisture recycling was found although the identification of the source areas needs further investigation. The upper-middle catchment area is formed by fractured shale/limestone on top of compact sandstone. The lower catchment area is comprised of an alluvial unit of about 15 m thickness overlaying a fractured shale unit. Two major groundwater flow systems were identified: one deep in the shale unit, recharged in the upper-middle catchment area; and one shallow, flowing in the alluvium unit and recharged locally in the lower catchment area.

Chapter 3 examines the hydrological and geomorphological controls on the water balance of the mangrove forest (0.2 km^2) during the dry period. The used multi-methods approach combined hydrology, hydrochemistry and geophysics. Precipitation is the main freshwater input. Tidal sand ridges are the key geomorphologic features which allowed an increase in water storage of 351 m^3 d^{-1} during a 22 day period. Large precipitation events cause breaking of the sand ridges by excess water, suddenly emptying the system. Grey water and pit latrines from the nearby town influence shallow groundwater quality, but also provide nutrients for the mangrove forest. Refreshening and salinization processes are controlled by the general groundwater flow direction. Hydraulic and hydrochemical influence of seawater on coastal piezometers seems to be controlled by the elevation of the water table and the tidal amplitude. All these conditions control forest subsistence during the dry season, which is essential for the mangrove forest to provide ecological and economic benefits such as protection against flooding, habitat for numerous species, and tourist attractions.

Chapter 4 analyzes the climatic water balance for the catchment for the period of 2010-2013, along with runoff components based on hydrograph separation. Hydrometry, geological characterization and hydrochemical and isotopic tracers (3–components hydrograph separation) were used. The climatic water balance was estimated for 2010/11, 2011/12 and 2012/13 with net values of 811 mm year^{-1}, 782 mm year^{-1} and -447 mm year^{-1}, respectively. Runoff components were studied at different spatial and temporal scales, demonstrating that different sources and temporal contributions are controlled by dominant landscape elements and antecedent rainfall. In forested sub-catchments, permeable soils, stratigraphy and steep slopes favor subsurface stormflow generation contributing 50% and 53% to total discharge. At catchment scale, landscape elements such as smooth slopes, wide valleys, deeper soils and water table allow groundwater recharge during rainfall events. Groundwater dominates the hydrograph (50% of total discharge) under dry prior conditions. However, under wet prior conditions low soil infiltration capacity generates a larger surface runoff component (42%) which dominates total discharge. The results show that forested areas are important to reduce surface runoff and likely soil degradation which is relevant for the design of water management plans.

Chapter 5 discusses field scale experiments using bacterial DNA as natural hydrological tracers. It reports a field scale (11000 m) test of natural occurring bacterial DNA as a tracer during rainfall–runoff events. Synoptic sampling throughout the catchment was performed to determine background bacterial DNA content. Inhibitory substances present in surface runoff contributions to stream water affect DNA amplification during quantitative Polymerase Chain Reaction (qPCR). This is observed in the inhibition of qPCR for surface water samples during the rainy season. Groundwater samples collected in this period were not inhibited, but bacterial content decreased; probably due to dilution from local precipitation. Sample dilution combined with the use of bovine serum albumina (BSA) in the qPCR mix solves the inhibition issue. However, the optimal concentration of BSA should be further investigated. The DNA harvesting method used *in situ* was successful. Nonetheless, DNA losses during the pre-filtration step have to be evaluated. This is a promising technique for hydrological research, but more field scale experiments are required to use bacterial DNA to investigate rainfall–runoff processes in a quantitative way. DNA recovery and qPCR inhibition in runoff samples have to be addressed in future works. Future experimentation should include areas with different soil types.

Chapter 6 looks into seasonal river–aquifer interactions in this area which is dominated by tidal sand ridges. The effect of stream stage fluctuations on river–aquifer flows and pressure propagation in the adjacent aquifer was investigated analyzing high temporal resolution hydraulic head data and applying a numerical model (HYDRUS 2D). Tidal sand ridges at the river outlet control the flow direction between the river and the aquifer. Surface water accumulation caused by these features induces aquifer recharge from the river. Simulations show groundwater recharge up to 0.2 m^3 h^{-1} per unit length of river cross section. Rupture of the sand ridges due to overtopping river flows causes a sudden shift in the direction of flow between the river and the aquifer. Groundwater exfiltration reached 0.08 m^3 h^{-1} immediately after the rupture of the sand ridges. Simulated bank storage flows are between 0.004–0.06 m^3 h^{-1}. These estimates are also supported by the narrow hysteresis loops between hydraulic heads and river stage which indicate small changes in groundwater levels. The aquifer behaves as confined, rapidly transmitting pressure changes caused by the river stage fluctuations. However, the pressure wave is attenuated with increasing distance from the river. Therefore, we concluded that a dynamic pressure wave is the mechanism responsible for the observed aquifer responses. Pressure variation observations and numerical groundwater modeling are useful to examine river–aquifer interactions and should be coupled in the future with chemical data to further improve process understanding.

Finally, Chapter 7 provides overall conclusions on the geomorphologic and hydro-climatic controls on catchment response, implications for sustainable water resources management, the link between hydrological processes and ecosystems; and implications for other similar catchments in this region. This is followed by recommendations for future research. The main conclusions are summarized here:

- Catchment structure (topography, geology and land use) controls surface and subsurface runoff generation:

  Stratigraphy and topography determine two major groundwater flow systems: one regional system located in the shale/limestone unit and one local system, located in the clay/alluvium unit.

  Tidal sand ridges prevent river discharge into the ocean, inducing surface water accumulation in the lower catchment area. This yields a positive water balance which enables the subsistence of the mangrove forest during dry periods; and induces aquifer recharge by increasing river stage during dry and wet periods.

  Catchment scale, stratigraphy, slopes, soils and land use control surface runoff generation processes. Steep slopes, forested hillslopes, permeable soils and less permeable shale layers favor subsurface stormflow in the upstream sub-catchments. At the catchment level, a wider valley, thicker alluvial deposits and smooth slope favor major contributions of groundwater.

- High temporal and spatial variability of precipitation, even for a relatively small catchment, affects availability of water resources for specific ecosystems and humans; determines sources of surface runoff generation and induces changes in groundwater–surface water interactions.

- Sustainable water resources management must prevent drastic alterations in catchment structural characteristics such as forested areas and tidal sand ridges. Forested hillslopes in the upper catchment area are crucial to reduce surface runoff and likely soil erosion. Groundwater recharge occurs in these areas and travels downstream the catchment, where it sustains river baseflow during dry periods and can be also used for future touristic development. Sand ridges regulate river stage increases during rainfall events and also during dry periods, thus controlling river–aquifer interactions. Groundwater recharge from river water is crucial during dry periods, especially considering the dependence of the local community on shallow groundwater resources.

- The catchment response to the hydro-climatic and geomorphologic controls described, supports the mangrove ecosystem freshwater needs. Surface water accumulation by means of the sand ridges enables the ecosystem to function and survive through dry periods. The mangrove in turn, provides a natural defense against flooding, provides habitat for numerous species, provides socio-economic benefits as a touristic attraction and acts as a nutrient sink.

The outcome of this work is a contribution to the hydrological knowledge of poorly gauged catchment in humid tropics. It also provides scientific hydrological insights to support water resources management on the South Pacific Coast of Nicaragua, since the results are also applicable to similar catchments in the region. Future research should include long term hydro-climatic and water quality monitoring, effects of extreme events on river-aquifer interactions, investigation of moisture recycling in forested areas, hillslope processes and mangrove conservation.

Samenvatting

Hydrologisch onderzoek in de vochtige tropen is bijzonder uitdagend vanwege de sterk wisselende hydrologische omstandigheden. In deze regio's zijn er ook grote sociaal-economische spanningen, die veroorzaakt worden door de snelle toename van de bevolking en die leiden tot veranderingen in het landgebruik. Bovendien induceren de klimaatverandering en de variabiliteit ook veranderingen in het hydrologische regime. Desalniettemin is inzicht in de hydrologische processen in deze gebieden beperkt en is de overdracht van hydrologische kennis uit andere hydro-klimatologische gebieden naar tropische stroomgebieden met een hoge luchtvochtigheid niet altijd direct mogelijk vanwege de intrinsieke verschillen. Dit is vooral problematisch voor ontwikkelingslanden, waar het gebrek aan betrouwbare voorspellingen duurzaam beheer van watervoorraden verder bemoeilijken. Midden-Amerika, en Nicaragua in het bijzonder, is hier een goed voorbeeld van.

Het doel van het hier gepresenteerde onderzoek is om de processen te begrijpen die oppervlakte- en ondergrondse afvoer genereren in een beperkt bemeten kustelijk stroomgebied in Nicaragua onder vochtige tropische omstandigheden. Dit onderzoek richt zich specifiek op het identificeren van geomorfologische en hydro-klimatologische controle op de respons van stroomgebieden met verschillende ruimtelijke en temporele schalen; bestudeert het verband tussen de hydrologische processen en de gesteldheid van het ecosysteem (in dit geval mangrovebos); en analyseert de significantie van processen die afvoer genereren voor het beheer van waterbronnen.

Het studiegebied deelt de topografische, geologische en hydro-klimatologische kenmerken van andere stroomgebieden in het Stille Zuidzee-gebied van Nicaragua. Landgebruik omvat gewoonlijk bosbouw, landbouw en veehouderij. Mangrove-ecosystemen zijn kenmerkend voor deze stroomgebieden. De bevolking is voor de watervoorziening meestal afhankelijk van ondiep grondwater en sanitaire voorzieningen ontbreken. Echter, de Zuid-Pacifische kust van Nicaragua heeft een groot toeristisch potentieel en vastgoedontwikkeling verloopt snel. De toename van het toerisme en andere gerelateerde ontwikkelingen zal de druk op de watervoorraden in deze regio verder verhogen.

Dit proefschrift is als volgt opgebouwd: Hoofdstuk 1 geeft een overzicht van het waterbeheer in beperkt bemeten stroomgebieden en hydrologische uitdagingen in de vochtige tropen. Het geeft ook de situatie weer van de watervoorraden in Nicaragua, en het beschrijft het studiegebied. Hoofdstuk 2 onderzoekt grondwaterstromingssystemen met behulp van een combinatie van geofysische, hydrochemische en isotopische methoden. Elektrische weerstand tomografie (ERT) is toegepast langs een 4.4 km transect, parallel aan de hoofdstroom, en in vijf dwarsdoorsneden van de rivier, om fracturen te identificeren en om de geometrie van de aquifer te bepalen. Stabiele waterisotopen, chloride en silica werden geanalyseerd in monsters van waterbronnen, de rivier, waterputten en piëzometers tijdens het droge en natte seizoen van 2012. Er werd een indicatie voor een waterkringloop gevonden, hoewel de brongebieden nog verder onderzocht moet worden. Het bovenste en middelste stroomgebied wordt gevormd door gebroken leisteen / kalksteen met daar bovenop compacte zandsteen. Het onderste stroomgebied bestaat uit een alluviale eenheid van ongeveer 15 m dik dat een gebroken schalie-eenheid bedekt. Twee belangrijke grondwaterstromingssystemen werden geïdentificeerd: een diep systeem in de schalie-eenheid gevoed in het bovenste en middelste stroomgebied; en een ondiep systeem dat stroomt in de alluviumeenheid en lokaal gevoed wordt in het onderste stroomgebied.

Hoofdstuk 3 gaat in op de hydrologische en geomorfologische invloed op de waterbalans van het mangrove bos (0.2 km^2) tijdens de droge periode. Bij de gebruikte multi-methode-aanpak zijn hydrologie, hydrochemie en geofysica met elkaar gecombineerd. Neerslag is de belangrijkste aanvoer van zoetwater. Zandruggen zijn de belangrijkste geomorfologische kenmerken, die een toename van de waterberging van 351 m^3 d^{-1} gedurende een periode van 22 dagen mogelijk maakten. Hevige neerslag veroorzaakt het breken van de zandruggen door overtollig water, waardoor het systeem plotseling leeg raakt. Grijswater en beerputten uit de nabijgelegen stad beïnvloeden de kwaliteit van het ondiepe grondwater, maar zorgen ook voor extra voedingsstoffen voor het mangrovebos. Verversings- en verziltingsprocessen worden beheerst door de richting van de gewone grondwaterstroming. De hydraulische en hydrochemische invloed van het zeewater op de piëzometers lijkt te worden beheerst door de verhoging van de grondwaterstand en de amplitude van het getij. Deze condities bepalen het voortbestaan van het het bos tijdens het droge seizoen, wat essentieel voor het bieden van potentiële ecologische en economische voordelen van het mangrovebos zoals bescherming tegen overstromingen, habitat voor tal van fauna en de attractie voor toeristisme.

Hoofdstuk 4 analyseert de klimatologische waterbalans voor het stroomgebied in de periode 2010-2013, samen met de componenten in de waterstromen op basis van hydrograafscheiding. Hydrometrie, geologische karakterisering en hydrochemische en isotooptracers (3-componenten hydrograafscheiding) werden gebruikt. De klimatologische waterbalans is geschat voor 2010/11, 2011/12 en 2012/13 met een netto waarde van 811 mm jaar^{-1}, 782 mm jaar^{-1} en -447 mm jaar^{-1}, respectievelijk. Afvoercomponenten zijn bestudeerd op verschillende ruimtelijke en temporele schalen, waaruit is gebleken dat verschillende bronnen en temporele bijdragen worden beheerst door dominante landschapselementen en antecedentneerslag. In doorlatende gronden van bosrijke subbekkens bevorderen stratigrafie en steile hellingen de generatie van ondergrondse afvoer, die voor respectievelijk 50% en 53% bijdragen aan de totale afvoer. Op het niveau van stroomgebied maken landschapselementen zoals gladde hellingen, brede valleien, diepe bodem en grondwater, aanvulling van het grondwater tijdens regenval mogelijk. Grondwater domineert de hydrografie (50% van de totale afvoer) onder voorafgaand droge omstandigheden. Echter, een lage bodeminfiltratiecapaciteit genereert een grotere oppervlakte-afvoercomponent (42%) onder voorafgaand natte omstandigheden die de totale afvoer domineert. De resultaten laten zien dat beboste gebieden belangrijk zijn voor de vermindering van oppervlakteafvoer en daarmee dus voor een vermindering van de aantasting van de bodem, wat relevant is voor het ontwerp van de waterbeheerplannen.

Hoofdstuk 5 bespreekt veldexperimenten, waarin bacterieel DNA als natuurlijke hydrologische tracer gebruikt wordt. Het beschrijft een veldtest met een schaal van 11,000 m met natuurlijk voorkomend bacterieel DNA als een tracer tijdens afvoer van neerslag. Synoptische bemonstering werd in het hele stroomgebied uitgevoerd om het achtergrondgehalte te bepalen. Remmende stoffen die aanwezig zijn in de afvoer beïnvloeden de DNA amplificatie tijdens de qPCR. Dit wordt waargenomen in de inhibitie van de oppervlaktewatermonsters tijdens het regenseizoen. Grondwatermonsters die in deze periode verzameld waren, werden niet geïnhibeerd, maar het bacteriële gehalte daalde waarschijnlijk als gevolg van verdunning door lokale neerslag. Monsterverdunning gecombineerd met het gebruik van bovine serum albumina (BSA) in de qPCR mix verhielp het inhibitieprobleem. De optimale concentratie van BSA moet echter nader onderzocht worden. De gebruikte *in situ* DNA-verzamel methode was succesvol. Toch moeten verliezen van DNA tijdens de prefiltratiestap geëvalueerd worden. Meer veldexperimenten zijn nodig naar het inzetten van bacterieel DNA om op een kwantitatieve manier neerslag- en afvoerprocessen te onderzoeken. Terugwinning van DNA en de inhibitie van qPCR in afvoermonsters zullen in toekomstig onderzoek aan de orde moeten komen. Toekomstig experimenteel onderzoek zal gebieden met verschillende grondsoorten moeten omvatten.

Hoofdstuk 6 gaat in op seizoensgebonden rivier-aquifer interacties in dit gebied wat wordt gedomineerd door de kustruggen. Het effect van schommelingen in de stroomfase op de rivier-aquifer stromen en de drukpropagatie in de aangrenzende aquifer werd onderzocht door het analyseren van stijghoogtedata met een hoge tijdsresolutie en het toepassen van een numeriek model (HYDRUS 2D). Zandruggen aan de riviermond beheersen de stroomrichting tussen de rivier en de aquifer. Accumulatie van oppervlaktewater veroorzaakt door deze functies en het induceert het voeden van de aquifer door de rivier. Simulaties tonen grondwateraanvulling aan tot 0.2 m^3 h^{-1} per lengte-eenheid van de rivierdwarsdoorsnede. Het breken van de zandruggen als gevolg van overslaande rivierstromen veroorzaakt een plotselinge verschuiving in de richting van de stroming tussen de rivier en de aquifer. Grondwater exfiltreerde met een snelheid van 0.08 m^3 h^{-1} onmiddellijk na de breuk van de zandrichels. Gesimuleerde bankopslagstromen hebben een snelheid tussen 0.004 en 0.06 m^3 h^{-1}. Deze schattingen worden ook ondersteund door de smalle hystereselussen tussen de stijghoogten en het rivierniveau. De aquifer gedraagt zich als begrensde, snel overgedragen drukveranderingen veroorzaakt door de rivierniveauschommelingen. Echter, de drukgolf verzwakt naarmate de afstand tot de rivier toeneemt. Daarom kon er geconcludeerd worden dat een dynamische drukgolf het mechanisme is dat verantwoordelijk is voor de waargenomen aquifer respons. Drukvariatiewaarnemingen en numerieke grondwatersysteemmodellering zijn bruikbaar om rivier-aquifer interacties te onderzoeken en deze moeten in de toekomst gekoppeld worden met chemische gegevens om het begrip van het proces verder te verbeteren .

Tot slot geeft Hoofdstuk 7 algemene conclusies over de geomorfologische en hydro-klimatologische effecten op de respons van het stroomgebied, implicaties voor duurzaam waterbeheer en het verband tussen hydrologische processen en ecosystemen. Dit wordt gevolgd door aanbevelingen voor toekomstig onderzoek. De belangrijkste conclusies worden hier samengevat:

- De structuur van het stroomgebied (topografie, geologie en landgebruik) beheerst de generatie van bovengrondse en ondergrondse stroming:

  Stratigrafie en topografie bepalen twee grote grondwaterstromingssystemen: een regionaal systeem gelegen in de leisteen- / kalksteenlaag en een lokaal systeem gelegen in de klei / alluviumlaag. Zandruggen belemmeren de doorstroom vanuit de rivier richting de oceaan, wat leidt tot accumulatie van oppervlaktewater in het onderste stroomgebied. Dit veroorzaakt een positieve water balans dat het onderhoud van het mangrovebos in droge perioden mogelijk maakt; en induceert voeding van de aquifer door het verhogen van het rivier niveau tijdens droge en natte periodes. Bekkenschaal, stratigrafie, hellingen, bodem en landgebruik beheersen de generatie van processen vooroppervlakte afvoer. Steile hellingen, beboste heuvelhellingen, doorlatende bodem en minder doorlatende schalielagen bevorderen ondergrondse afvoer in de stroomopwaartse substroomgebieden. Op het niveau van stroomgebieden bevorderen een breder dal, dikkere alluviale afzettingen en gladde hellingen een grote bijdrage van het grondwater.

- Hoge temporele en ruimtelijke variabiliteit van neerslag, zelfs voor een relatief klein stroomgebied, beïnvloedt beschikbaarheid van water voor specifieke ecosystemen en de mens; bepaalt bronnen voor generatie van oppervlaktewaterafvoer en leidt tot veranderingen in de grondwater-oppervlaktewater interacties.

- Duurzaam waterbeheer moet drastische veranderingen in de structurele kenmerken van het stroomgebied, gedefinieerd door beboste gebieden en zandruggen voorkomen.

Beboste heuvelhellingen in het bovenste stroomgebied zijn cruciaal om oppervlakte afvoer en dus bodemerosie te verminderen. Aanvulling van het grondwater treedt op in deze gebieden en beweegt zich stroomafwaarts in het stroomgebied, waar het de basis rivier stroom in stand houdt in droge perioden en kan ook worden gebruikt voor toekomstige toeristische ontwikkelingen. Zandruggen reguleren stijgingen in het rivier niveau tijdens regenval en ook in droge perioden, waardoor ze rivier-aquifer interacties beheersen. Grondwateraanvulling met rivierwater is cruciaal tijdens droge periodes, zeker gezien de afhankelijkheid van de lokale gemeenschap op ondiepe grondwaterbronnen.

- De beschreven respons van het stroomgebied op de hydro-klimatologische en geomorfologische controle karakteristieken, ondersteunt de zoetwater behoeften van het mangrove ecosysteem. Accumulatie van oppervlakte water door de zandruggen maakt het mogelijk dat het ecosysteem functioneert en de droogte overleeft. De mangrove op zijn beurt, zorgt voor een natuurlijke bescherming tegen overstromingen, biedt leefgebied voor tal van diersoorten, biedt sociaaleconomische voordelen als een toeristische attractie, en fungeert als een opslag voor nutriënten. Het resultaat van dit werk is een bijdrage aan de hydrologische kennis van slecht gemeten stroomgebied in vochtige tropen. Het biedt ook de wetenschappelijke hydrologische inzichten, die het waterbeheer ondersteunen op de Zuid-Pacifische kust van Nicaragua.

Toekomstig onderzoek zou moeten omvatten: lange termijn hydro-klimatologische en waterkwaliteit monitoring, de effecten van extreme gebeurtenissen op de rivier-aquifer interacties; het onderzoek van vochtrecycling in beboste gebieden, processen van heuvelhellingen en behoud van mangrove.

Resumen

La investigación hidrológica en los trópicos húmedos es particularmente desafiante debido a la alta variabilidad de las condiciones hidrológicas. Estas regiones sufren además grandes estreses socio-económicos debido al rápido aumento de la población, lo cual conlleva cambios en el uso del suelo. Además, la variabilidad y el cambio climático inducen cambios en el régimen hidrológico. Sin embargo, el entendimiento de los procesos hidrológicos en estas áreas es limitado y la transferencia de conocimiento hidrológico de otras regiones hidro-climáticas a cuencas en los trópicos húmedos puede ser difícil debido a sus diferencias intrínsecas. Esto es especialmente problemático para países en desarrollo, donde las limitaciones para producir predicciones confiables impide la gestión sostenible de los recursos hídricos. Centro América, y Nicaragua en particular, son buenos ejemplos de estas regiones.

El objetivo de esta investigación es entender los procesos de generación de escorrentía superficial y subterránea en una cuenca costera de Nicaragua pobremente aforada bajo condiciones de trópico húmedo. Específicamente, esta investigación se enfoca en identificar los controles geomorfológicos e hidro-climáticos en la respuesta de la cuenca a diferentes escalas espacio-temporales; estudia la relación entre procesos hidrológicos y condiciones de los ecosistemas (*i.e.* bosques de mangle); y analiza la significancia de los procesos de generación de escorrentía para la gestión sostenible de los recursos hídricos.

El área de estudio comparte las características topográficas, geológicas e hidro-climáticas de otras cuencas del Pacífico Sur de Nicaragua. El uso del suelo incluye bosques, agricultura y pastos. Los ecosistemas de manglar se encuentran típicamente en estas cuencas. La población depende principalmente del agua subterránea somera para suministro de agua y no existen sistemas de saneamiento. Sin embargo, la costa del Pacífico Sur de Nicaragua posee un enorme potencial turístico y el desarrollo inmobiliario está dándose rápidamente. El aumento en el turismo y otros desarrollos relacionados aumentara aún más el estrés sobre los recursos hídricos de esta región.

Esta tesis está organizada de la siguiente manera: el Capítulo 1 proporciona una descripción de la gestión de los recursos hídricos en cuencas pobremente aforadas y los desafíos hidrológicos en los trópicos húmedos. Asimismo presenta una visión general la situación de los recursos hídricos en Nicaragua y describe el área de estudio. El Capítulo 2 investiga los sistemas de flujo de agua subterránea usando una combinación de geofísica, hidroquímica y métodos isotópicos. Se aplicó Tomografía de Resistividad Eléctrica (TRE) a lo largo de un transecto de 4.4 km paralelo al río principal y en cinco secciones transversales, para identificar fracturas y determinar la geometría del acuífero. Se analizaron isótopos estables del agua, cloruros y sílica en muestras de manantiales, río, pozos y piezómetros para las épocas seca y lluviosa del 2012. Se encontraron indicios de reciclaje de humedad aunque la identificación del origen de esta humedad requiere más investigación. La parte alta-media de la cuenca está formada por lutitas/calizas superpuestas sobre areniscas compactas. La parte baja de la cuenca está formada por una unidad aluvial de aproximadamente 15 m de espesor, que sobreyace una unidad de lutitas fracturadas. Se identificaron dos sistemas de flujo principales: un sistema profundo en la unidad de lutitas, recargado en la parte alta-media de la cuenca; y otro sistema somero, que fluye en la unidad aluvial y es recargado localmente en la parte baja de la cuenca.

El Capítulo 3 examina los controles hidrológicos y geomorfológicos en el balance hídrico de un bosque de mangle (0.2 km^2) durante la época seca. Se usó un enfoque multi-método que combina hidrometría, hidroquímica y geofísica. La precipitación es el principal contribuyente de agua dulce. Las crestas de arena producto de las mareas son las características geomorfológicas principales que permiten el aumento en el almacenamiento de agua de 351 m^3 d^{-1} durante un periodo de 22 días. Eventos grandes de precipitación causan la ruptura de las crestas de arena debido al exceso de agua, vaciando instantáneamente el sistema. Las aguas grises y las letrinas del pueblo cercano influencian la calidad del agua subterránea somera, pero también proporcionan nutrientes para el bosque de mangle. Los procesos de salinización y refrescamiento son controlados por la dirección general del agua subterránea. La influencia hidráulica e hidroquímica del agua de mar en los piezómetros costeros parece estar controlada por la elevación del nivel freático y la amplitud de la marea. Estas condiciones controlan la subsistencia del bosque de mangle durante la época seca, lo cual es esencial para que el bosque proporcione beneficios ecológicos y económicos tales como protección contra inundaciones, hábitat para numerosas especies y atracción turística.

El Capítulo 5 discute experimentos a escala de campo usando ADN de bacterias como un trazador hidrológico natural. Reporta experimentos a escala de campo (11,000 m) usando ADN de bacterias naturales como trazador durante eventos de precipitación–escorrentía. Se realizó un muestreo sinóptico en la cuenca para determinar el contenido normal de ADN bacterial. Substancias inhibitorias presentes en las contribuciones de escorrentía superficial al río afectan la amplificación del ADN durante el proceso de reacción en cadena de la polimerasa (qPCR). Esto se observa en la inhibición de qPCR en las muestras de agua superficial durante la época lluviosa. Las muestras de agua subterránea colectadas durante este periodo no mostraron inhibición, pero el contenido de ADN bacterial disminuyó probablemente debido a dilución causada por la precipitación local. La dilución de las muestras combinada con el uso de albumina de suero bovino (BSA) en la mezcla de qPCR resuelve el problema de inhibición. Sin embargo, la concentración óptima de BSA debe ser investigada con mayor detalle. El método de cosecha de ADN usado *in situ* fue exitoso. No obstante, las pérdidas de ADN durante el proceso de pre-filtración deben ser evaluadas. Esta es una técnica promisoria para investigaciones hidrológicas, pero se necesitan más experimentos a escala de campo para poder usar el ADN bacterial para investigar de manera cuantitativa los procesos de precipitación–escorrentía. La recuperación de ADN y la inhibición durante qPCR en las muestras de escorrentía deben ser investigadas en trabajos futuros. Además, los experimentos futuros deben incluir áreas con diferentes tipos de suelo.

El Capítulo 6 analiza las interacciones estacionales entre el río y el acuífero en el área dominada por las crestas de arena en la costa. El efecto de las fluctuaciones del nivel del río en los flujos entre el río y el acuífero y la propagación de presión en el acuífero adyacente fue investigada usando datos de cargas hidráulicas de alta resolución temporal y aplicando un modelo numérico (HYDRUS 2D). Las crestas de arena originadas por la marea controlan la dirección del flujo entre el río y el acuífero. La acumulación de agua superficial causada por las barras de arena inducen recarga del acuífero desde el río. Las simulaciones muestran recarga de agua subterránea de hasta 0.2 m^3 h^{-1} por unidad de longitud de sección transversal del río. La ruptura de las barras de arena a causa de la acumulación de agua causa un súbito cambio en la dirección del flujo entre el río y el acuífero. La descarga de agua subterránea hacia el río alcanza 0.08 m^3 h^{-1} inmediatamente después de la ruptura de las barras de arena. Los flujos de almacenamiento en los bancos del río están entre 0.004-0.06 m^3 h^{-1}. Estas estimaciones son también soportadas por la forma cerrada de los gráficos de histéresis entre las cargas hidráulicas y los niveles del río, que indican cambios pequeños en los niveles de agua subterránea. El acuífero se comporta como confinado, transmitiendo rápidamente los cambios de presión producidos por las fluctuaciones en el nivel del río. Sin embrago, la onda de presión es atenuada al alejarse del río. Por lo tanto, concluimos que el mecanismo responsable de la respuesta del acuífero es una onda de presión dinámica. Las observaciones de variación de presión y el modelo numérico son útiles para examinar las interacciones entre el río y el acuífero y deben ser acopladas en el futuro con datos químicos para mejorar más el entendimiento de estos procesos.

Finalmente, el Capítulo 7 presenta las conclusiones generales sobre los controles geomorfológicos e hidro-climáticos sobre la respuesta de la cuenca, implicaciones en la gestión sostenible de los recursos hídricos, la conexión entre los procesos hidrológicos y los ecosistemas; e implicaciones de estos resultados para otras cuencas de la región. Esto es seguido por recomendaciones para investigaciones futuras. Las conclusiones principales son resumidas aquí:

- La estructura de la cuenca (topografía, geología y uso del suelo) controlan la generación de escorrentía superficial y subterránea.

  La estratigrafía y la topografía determina dos sistemas de flujo subterráneo principales: un sistema regional localizado en la unidad de lutitas/calizas y otro sistema local, localizado en la unidad aluvial.

  Las crestas de arena en la costa previene la descarga del río al océano, induciendo acumulación de agua superficial en la parte baja de la cuenca. Esto produce un balance hídrico positivo el cual permite la subsistencia del bosque de mangle durante periodos secos; e induce recarga desde el río al acuífero al incrementar el nivel del río durante los periodos seco y lluvioso.

  La escala de la cuenca, la estratigrafía, las pendientes, suelos y el uso del suelo controlan los procesos de generación de escorrentía superficial. Las fuertes pendientes, las laderas boscosas, los suelos permeables y la capa de lutita menos permeable favorecen el flujo sub-superficial en las sub-cuencas ubicadas en la parte alta de la cuenca. A escala de cuenca, la amplitud del valle, los depósitos aluviales y la suave pendiente favorecen una mayor contribución del agua subterránea.

- La alta variabilidad temporal y espacial de la precipitación, aun para una cuenca relativamente pequeña, afecta la disponibilidad de recursos hídricos para ecosistemas específicos y para los seres humanos; determina las fuentes de generación de escorrentía e induce cambios en las interacciones río-acuífero. subterránea.

- La gestión sostenible de recursos hídricos debe prevenir alteraciones drásticas en las características estructurales de la cuenca definas por las áreas boscosas y las barras de arena en la costa. Las laderas boscosas en la parte alta de la cuenca son cruciales para reducir la escorrentía superficial y por lo tanto la erosión del suelo. La recarga de agua subterránea ocurre en estas áreas y fluye aguas abajo, donde sostiene el flujo base del río durante periodos secos y además puede ser utilizado para el desarrollo turístico. Las barras de arena regulan el incremento del nivel del río durante eventos de precipitación y además durante periodos secos, controlando así las interacciones río-acuífero. La recarga de agua subterránea desde el río es crucial durante periodos secos, especialmente considerando la dependencia de las comunidades locales en los recursos subterráneos someros.

- La respuesta de la cuenca a los controles hidro-climáticos y geomorfológicos abastecen las necesidades de agua dulce del ecosistema de mangle. La acumulación de agua superficial favorecida por las crestas de arena habilitan al ecosistema para funcionar y sobrevivir durante periodos secos. El bosque de mangle, a cambio, proporciona beneficios socio-económicos como atractivo turístico, protección contra inundaciones y captador de nutrientes.

El resultado de este trabajo es una contribución al conocimiento hidrológico de cuencas pobremente aforadas en los trópicos húmedos. Además proporciona información científica hidrológica para apoyar la gestión de los recursos hídricos en la costa del Pacífico Sur de Nicaragua, ya que los resultados son aplicables a cuencas similares en la región. Las investigaciones futuras deberán incluir monitoreo hidro-climáticos y de calidad de agua a largo plazo, los efectos de eventos extremos en las interacciones entre el rio y el acuífero, investigar el reciclaje de humedad en las áreas boscosas, procesos hidrológicos de laderas y la conservación del bosque de mangle.

List of Symbols and Acronyms

Symbols:

CEC	Cation exchange capacity	$[N\ M^{-1}]$
Ct	Threshold cycle	$[-]$
EC	Electrical conductivity	$[L^{-1}M^{-2}T^{-3}I^{2}\ L^{-1}]$
f_{sea}	Fraction of sea water	$[\%]$
h	Hydraulic head	$[L]$
hp	Pressure head	$[L]$
Ks	Saturated hydraulic conductivity	$[L\ T^{-1}]$
MAE	Mean Absolute Error	$[L]$
n	Porosity	$[-]$
Q	Discharge	$[L^{3}\ T^{-1}]$
R^{2}	Correlation Coefficient	$[-]$
RMSE	Root Mean Square Error	$[-]$
T	Temperature	$[\theta]$
θs	Saturated soil water content	$[L^{3}\ L^{-3}]$
θr	Residual soil water content	$[L^{3}\ L^{-3}]$
α	Coefficient in the soil water retention function	$[L^{-1}]$

Acronyms:

ANA	National Water Authority
BSA	Bovine Serum Albumina
CIRA	Nicaraguan Aquatic Resources Research Center
DEM	Digital Elevation Model
DEPC	Diethyl Pyrocarbonate
DNA	Deoxyribonucleic acid
ERT	Electrical Resistivity Tomography
GMWL	Global Meteoric Water Line
GWP	Global Water Partnership
INETER	Nicaraguan Institute of Territorial Studies
IPCC	Intergovernmental Panel on Climate Change
LMWL	Local Meteoric Water Line
LT	Local Time
m asl	meters above sea level
m bgl	meters below ground level

MSD	Midsummer drought
NA	Nucleic acid
PUB	Prediction in Ungauged Basins
qPCR	Quantitative Polymerase Chain Reaction
SICA	Central American Integration System
SINAPRED	National System for Disaster Prevention
SRTM	Shuttle Radar Topography Mission
UNA	National University of Agriculture

Table of Contents

Chapter 1

Introduction

1.1. Water resources management in poorly gauged tropical catchments

Catchment hydrology deeply links with water resources management to ensure life and ecosystem sustainability (Bonell and Bruijnzeel 2004, Uhlenbrook 2006). Sustainable water management should guarantee water for human life, preservation and sustainable use of ecosystems and also minimize the impact of natural water related hazards. Water management decisions require continuous information of water resources and reliable predictions of hydrological responses to optimally integrate social, economic and ecological perspectives. However, many catchments in the world are ungauged or poorly gauged and the capability to make these predictions is limited (Sivapalan et al. 2003).

In view of the lack of observed data, hydrological behavior can be inferred from catchment physical and climatic characteristics or from hydrologically similar gauged catchments (Singh et al. 2014). This is the topic of contemporary research. Generalization of knowledge from gauged to ungauged (or poorly gauged) catchments should rely on a catchment classification system to help determine similarities and differences (Sivapalan 2006). In this context, the decade on Prediction of Ungauged Basins (PUB) initiative emerged (Sivapalan et al. 2003). PUB aimed to improve the capabilities of the scientific community to make predictions in ungauged basins. Among the outputs of the PUB initiative are the progress made in linking catchment characteristics to catchment function. This lead to advances in catchment classification schemes (e.g. Gaál et al. 2012) , similarity frameworks (e.g. McDonnell and Woods 2004, Savenije 2010) and model regionalization methods for transferring knowledge from gauged to ungauged catchments and improve predictions in the latter (e.g. Lyon et al. 2012) . Also, the links between catchment form and function lead to the recognition that hydrology is as an integral part of the ecosystem, and that a holistic approach would improve understanding of catchment organization and function (Hrachowitz et al. 2013).

Kirchner (2003) and Sivapalan (2006) state that the dominance of small-scale theories in catchment hydrology limits its ability to explain catchment behavior. Therefore, a unifying theory that explains different processes at different spatial and temporal scales, as well as across different hydro-climatic regions is necessary. This theory requires a catchment classification system to help determine similarities and differences between catchments and identify useful patterns (Sivapalan 2006). Such a catchment classification system should relate catchment structure (*e.g.* geology, topography, pedology and land use) and hydro-climatic characteristics to the catchment response. This response includes partition, storage and release of water (Wagener et al. 2007) and should include surface and groundwater flows, as well as residence times, water age and water chemistry at different spatial and temporal scales (Sivapalan 2006).

Increasing population around the world is causing land use changes from natural vegetation to agriculture, human settlements or industries. This triggers changes in surface runoff, groundwater recharge and flow, and water quality (Sivapalan et al. 2003) . Climate change and climate variability and other global changes also induce changes in the hydrological regime. Developing countries usually suffer significantly the impacts of climate and land use change, and they are also the regions with less hydrologic data and monitoring networks. This combination leads to depletion of water resources and ecosystem degradation (Sivapalan et al. 2003).

PUB proposed process studies and field experiments worldwide for theory development and model improvement. Still, improvements in hydrological predictions during the last decade have been mostly done in gauged, rather than in ungauged catchments. This is especially problematic for developing countries, where limitations to produce reliable predictions affect the ability to manage water resources (Hrachowitz et al. 2013). Thus, comparative hydrology across different hydro-climatic regions is necessary as it may allow identification of controls on transfer of hydrological parameters from gauged to ungauged catchments (Singh et al. 2014).

Hydrological process studies not only must consider the complex interactions between water, land use, soils, atmosphere and society, but also the interrelation between surface water and groundwater within the hydrological continuum (Alley et al. 2006, Savenije 2009, Winter et al. 2003, Hrachowitz et al. 2013, Sivapalan et al. 2003, Uhlenbrook 2006). These interactions are essential in understanding issues such as water supply, water quality and aquatic ecosystems (Alley et al. 2006, Sophocleous 2002, McClain et al. 2012) and, therefore, for the sustainable management of water resources. Failing to consider the interactions between surface and subsurface water resources may lead to false estimations and misconceptions and poor water resources management decisions.

Although the hydrological impact of forests has been widely investigated (e.g. Bruijnzeel, 2011) , the differences in climatological, pedological and physiological conditions between catchments, cause different hydrological responses to land use changes (Andréassian 2004). The effect of land use changes in the water fluxes in Central America remains unclear (Kaimowitz 2004). Simulation models have been focused on large regions where a large portion of rainfall comes from evaporation within the region, such as the case of the Amazonas basin. However, this is not the case for Central America where precipitation is largely of oceanic origin (Magaña et al. 1999). Research in Central America shows it is unlikely that land use changes over a few thousand of kilometers changes precipitation patterns (Bruijnzeel 2004a). However, at smaller scales, land use changes which reduce soil infiltration increase surface runoff and flow peaks. Forest clearing also increase water yield by reducing evaporation. However, in tropical cloud montane forests, the opposite may occur, since the forest can capture and recycle moisture (Bruijnzeel 2004b). Furthermore, infiltration favored by tree roots will replenish groundwater resources (Kaimowitz 2004).

These effects of land use change stress the need to take actions against the degradation of Central American catchments. In 1998, Hurricane Mitch caused 9000 victims and US$6 billion in damage in Central America. After the disaster, international cooperation and government agencies attributed the magnification of the damage to deforestation. This caused a surge in initiatives to address this problem by focusing on reforestation, soil conservation and civil defense (Kaimowitz 2004). Nevertheless, most catchment management projects gave insufficient attention to research and monitoring and were more guided by pre-conceived ideas, instead of trying to learn from research and design proper management strategies. This example highlights the need for water resources management in Central America needs to evolve from immediate crisis response to long term monitoring and planning (Kaimowitz 2004).

1.2. Hydrological challenges in humid tropics

Humid tropics are located 25° north and south of the Equator and include areas where precipitation exceeds evaporation at least 270 days per year (Wohl et al. 2012). These regions cover one fifth of the world's land surface and produce the largest amount of runoff. They also suffer from the greatest land cover change by forest clearing (FAO 2010). Hydrology of humid tropics differs from other world regions in the higher energy input in the form of water vapor fluxes, more intense precipitation, rapid weathering of inorganic and organic material and rapid movement of large volumes of sediment and water (Wohl et al. 2012). Furthermore, soil characteristics in humid tropical climates may differ from temperate climate soils because of differences in climate, flora and fauna (Minasny and Hartemink 2011). Additionally, the rate of human induced changes is faster because of population growth and socio-economic stresses (Wohl et al. 2012). Nevertheless, understanding of hydrological processes in this hydro-climatic regions is limited and transfer of hydrological knowledge from other hydro-climatic regions to tropical catchments may be challenging due to the difference in rainfall intensity and seasonality (Bonell 1993), in addition to the cyclical patterns of El Niño y La Niña (Bruijnzeel 2004a).

Prediction of water quantity and quality requires understanding of runoff generation processes (Bonell 1998); and according to Bonell and Bruijnzeel (2004) there has been relatively less research in tropical climate compared to the detailed studies carried out in temperate climates. Runoff generation processes in the tropics are expected to be different from temperate climates due to strong rainfall variability and seasonality. In addition, different soil types and land uses may cause differences between commonly studied temperate regions and poorly investigated tropical areas (Hugenschmidt et al. 2014). Figure 1.1 exemplifies the wider range of river discharge in tropical regions compared to temperate regions.

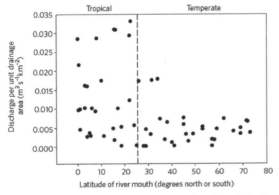

Figure 1.1 Average river discharge as a function of the location of river mouth for tropical and temperate regions from Wohl et al. (2012) based on data from Herschy and Fairbridge (1998)

High spatial and temporal climatic variability in the tropical regions may lead to periods and areas of highly variable hydrological conditions. Hydrological process are temporally and spatially highly variable and also governed by preferential flow (Uhlenbrook 2006). In addition to this variability, there are short moments and specific areas of exceptionally high hydrological activity. These specific time periods and areas are known as "hot moments" and "hot spots", respectively (McClain et al. 2003). Hot spots and hot moments vary in space and time, yielding different patterns for different processes. Additionally, it is necessary to consider scale differences between processes occurring in

3

surface water and groundwater systems. The processes occurring at a stream, especially those taking place at the hyporheic zone undergo seasonal and even daily cycles (Brunke and Gonser 1997, Alley et al. 2002), whereas groundwater processes have a much larger time scale of years, decades or more (Alley et al. 2002).

Rapid degradation and conversion of forested areas to other land uses in the humid tropics are altering the hydrological functioning of catchments (Bonell and Bruijnzeel 2004, Uhlenbrook 2007, Wohl et al. 2012, Elsenbeer 2001). Land use changes have significant influence on local and regional hydrology (Costa 2004). The nature of the impact of land use changes are grouped into soil impacts (e.g. Giertz *et al.* 2005, Diekkrüger and Hieppe 2012) and streamwater quality and quantity impacts (e.g. Chavez *et al.* 2008, Germer *et al.* 2009, Masese *et al.* 2014). These impacts can be summarized as increase in overland flow, erosion, sedimentation, peak flow, nutrient and chemical inputs; and decrease in water yields, baseflows, groundwater recharge and probably changes in precipitation regime at large scales (Aylward 2004).

Future alterations of the tropical hydrosphere will be driven by freshwater supply, agriculture and energy needs. These underlines the importance of understanding human–natural systems that will determine the future of the hydrologic cycle in these regions (Lele 2009). This is in line with the socio-hydrology concept discussed by Sivapalan et al. (2012), who state the need to observe, understand and predict the co-evolution of coupled human-water systems. The characteristics of humid tropical catchments, along with usually incomplete or non-existent hydrological records, represent a major challenge for hydrological investigations and adequate water resources management.

1.3.   Water resources in Nicaragua

1.3.1. The state of the art

Catchment areas in Central America are identified by numbers. On the Atlantic Coast they are assigned odd numbers, starting with 1 in Guatemala up to 121 in Panama. On the Pacific Coast they are assigned even numbers, starting with 2 in Guatemala up to 164 in Panama (PNUD and OMM 1972). Nicaragua has an extension of 130,000 km^2 and it is officially divided into 21 major catchment areas, out of which 13 drain to the Caribbean Sea and 8 drain to the Pacific Ocean. The catchment areas on the Caribbean side vary between 1,500 km^2 and 30,000 km^2, whereas on the Pacific side they range between 274 km^2 and 3,700 km^2. The River San Juan catchment is the largest and most important of the country. It covers about 30,000 km^2 and includes the two largest lakes of Central America: Lake Nicaragua, which has an approximate area of 8,200 km^2 and Lake Managua, which has an approximate extension of 1,040 km^2 (Castillo Hernández et al. 2006). There are also numerous crater lakes, some of which are used for water supply (*i.e.* Asososca) (Parello et al. 2008).

Hydrogeological mapping of the country started during the 1960s and 1970s (Castillo Hernández et al. 2006). Hydrogeological and hydrochemical maps (1:250,000) were elaborated by the Institute of Territorial Studies (Krasny and Hecht 1998). The central region of the country has also been mapped (INETER 2005). The hydrogeology of the Atlantic region has not been mapped yet but it is foreseen that INETER will carry out this task in the future (Castillo Hernández et al. 2006).

The most important groundwater resources of the country are found on the Pacific Coast. Most of them are unconfined aquifers recharged by precipitation. The most important are the Leon-Chinandega aquifer, located in the Northwestern part of the country in one of the most densely populated regions; and the Managua aquifer, located in the capital of the country. Although numerous studies have been performed in these aquifers (Corriols et al. 2009, Calderon and Bentley 2007, Moncrieff et al. 2008, Delgado Quezada 2003, Bethune et al. 1996, Cruz 1997, Johansson et al. 1999, Choza 2002), it is difficult to compare groundwater availability estimates due to disparities in methodologies and differences in the spatial and temporal density of the data used. Commonly in Nicaragua, an empirical method to estimate groundwater recharge from average monthly precipitation and soil infiltration capacity is used (Schosinsky and Losilla 2000). The method was developed based on the precipitation records from Costa Rica and although the authors state that it could possibly be applied in other areas of Central America with similar climatic characteristics; the uncertainty in the extrapolation of this method has not been assessed yet, but must be considered significant.

However, rough estimates of groundwater resources in the Pacific side indicate an availability of 3×10^9 million m^3 year^{-1} (Castillo Hernández et al. 2006). Groundwater reports in Nicaragua usually consider safe yield to be 50% of available groundwater resources, see for example the reviews by Castillo Hernández et al. (2006), GWP (2011) and Vammen et al. (2012). Nonetheless, the uncertainty of the groundwater fluxes estimates and the oversimplification in the application of the safe yield concept implies a threat to sustainable water resources use. The safe yield concept looks to avoid negative consequences of groundwater pumping such as depletion of stream flows, loss of wetlands and riparian ecosystems. However, it is not sufficient to consider only natural recharge to estimate the safe yield of an aquifer. The dynamic effects of groundwater extraction on the natural environment, the economy, and the society have to be considered too (Zhou 2009, Sophocleous 2007).

Additionally, there are many shortcomings in the estimation of catchment water balances. Evaporation is calculated through the modified Thornthwaite's method (Thornthwaite and Mather 1957), and the difference with precipitation is assumed to be equal to surface runoff. This approach does not consider interception explicitly, which can be especially important in forested catchments (Savenije 2004, Bruijnzeel 2004b). Discharge is usually not used as a direct input for the calculation, since monitoring stations are scarce and data is usually not available and often unreliable. Climatic records are also seldom complete, long-term, or available for the area of interest. Therefore, data extrapolation is a necessity and introduces associated uncertainties.

Although INETER runs a hydrometeorogical monitoring network for the country, the number and distribution of stations is not sufficient yet. Monitoring of groundwater levels was interrupted in 1979 and started again in 2003 for the main aquifers in the Pacific region (Castillo Hernández et al. 2006). However, monitoring is in most parts done manually and the temporal resolution is poor. In addition, there are 425 meteorological stations in the country, out of which 344 record only precipitation and only 30 are telemetric (webserver2.ineter.gob.ni/Direcciones/meteorologia/Red%20Meteorologica/antecedentes.htm). This situation highlights the need to extend and improve the characteristics of the hydrometeorological monitoring network in the country. This would improve the level of hydrological research and yield more reliable estimates of water resources in the country, allowing better management decisions.

Water quality studies are more prolific than physical hydrology research in Nicaragua. This is partly due to the significantly higher availability of human and financial resources. Water quality issues usually receive immediate attention from the public, government institutions and international cooperation agencies. Some examples of such studies are

Hassan et al. (1981), Lacayo et al. (1992), Calero et al. (1993), Carvalho et al. (1999), Castilho et al. (2000), Carvalho et al. (2003), Mendoza and Barmen (2006), Picado et al. (2010) and Scheibye et al. (2014). Among the major water quality issues in the country are diffuse contamination of surface water and groundwater by pesticides, contamination by arsenic from geologic origin, mercury pollution from artisanal mining and industry, contamination by waste waters, contamination from lixiviates from dumpsites, and sedimentation (GWP 2011, Castillo Hernández et al. 2006, Vammen et al. 2012).

Hydrochemical and isotope hydrology studies are also found, but the methods have been mostly applied to the major lakes and groundwater reservoirs of the country. Some examples are Payne and Yurtsever (1974), Araguas (1992), Plata et al. (1994), ARCAL XXI (1999), Parello et al. (2008) and INETER (2009). Applications to hydrological process studies are scarce. Some examples are Corrales and Delgado (2009), Calderon and Flores (2011), Calderon and Uhlenbrook (2014a) and Calderon et al. (in review).

1.3.2. Water resources legislation and development

Nicaragua has a National Water Resources Policy which was established in 2001. This instrument provides the general guidelines under which water resources management of the country should be conducted. The policy recognizes the finite nature of the water resources, as well as its economic, social, and environmental values. The policy has been under review and discussion since 2010 (Calderon 2010), in view of the fundamental socio-political changes undergone by the country. The discussion aims to update the policy accordingly to the new philosophical values of the government which are reflected in the National Plan of Human Development (2012) and also the General Law of National Waters (2007).

The General Law of National Waters of Nicaragua (Law No. 620) is notably very modern (2007). The law created the National Water Authority (ANA), which was established in 2010. This entity has the responsibility to 'manage the national water resources and their inherent goods' (National Assembly, 2007). Some of the most important duties of the ANA are the creation of a National Water Resources Plan and the creation of Water Management Plans for each of the 21 catchment areas of the country. The first one is a general plan to establish the priorities for water resources use and protection. The second are specific instruments to be applied in each catchment region and should be designed in a participatory way, including local and regional stakeholders. However, so far these duties have not been fulfilled, partly due to the shortage of qualified human resources and the limited financial resources.

Among the most notable features of the law, is the recognition of the human right to water, which primes over any other water use. The law also established the major lakes of the country, Lake Nicaragua and Lake Managua, as strategic water reservoirs and explicitly mandates their protection and conservation.

According to the review by Vammen et al. (2012), about 70% of the total water consumption is supplied by groundwater sources and the major water use in Nicaragua is agriculture (83%), followed by industrial use (14%) and domestic use (3%). However, this information does not include hydropower generation, which is becoming increasingly important for the county. Currently, the major renewable source is geothermal, followed by hydropower, wind, and biomass. Renewable energies account for 24% of the total energy demand. It is estimated that by 2020 they will cover 90% of the demand (http://www.cndc.org.ni/graficos/graficaGeneracion_Tipo_TReal.php). Although the water legislation requires the users to report the volumes of water utilized, there are still difficulties in enforcing this regulation. Therefore, the actual volumes of water used are not known.

Nicaragua is planning a US$ 40 billion project to construct a canal which connects the Caribbean Sea and the Pacific Ocean. All the proposed routes for the canal go across Lake Nicaragua and include the area of Rivas, in the Southwestern Pacific Coast. The project of the 'Interoceanic Grand Canal' aims to construct a multimodal and logistics center which will include not only the canal itself, but also major ports in the Caribbean and the Pacific Coast, free economic zones, a national railway and a new international airport (Oquist 2013). This project implies an enormous mobilization of natural, human, and financial resources and requires careful planning. The need to properly manage water resources for such a project requires among other things, a sound hydrological understanding of the catchments involved. This understanding can only be achieved by developing means of comprehensive hydrometeorological data collection and assigning human and financial resources to hydrological research.

1.3.3. Catchment hydrology and water resources management

Central America has a high spatio-temporal climatic variability despite its relatively small extend of about 520,000 km^2 (Westerberg et al. 2014). Water related disasters are frequent and include floodings and droughts with consequences such as loss of lives, crop damage, landslides, shortage in water supply and hydropower generation (George et al. 1998, Waylen and Sadí Laporte 1999). Although climate characteristics of the region have been studied (Amador et al. 2006, Durán-Quesada et al. 2012, Hastenrath 1967, Castillo Hernández et al. 2006, Hidalgo et al. 2013, Magaña et al. 1999), there are few peer-reviewed hydrological studies. Some examples are Genereux (2004), Mendoza et al. (2008), Harmon et al. (2009), Westerberg et al. (2010), Caballero et al. (2013) and Macinnis-Ng et al. (2014). The scarcity in literature is partly caused by limited access to adequate, complete, long term hydrometeorological data (Durán-Quesada et al. 2012, Westerberg et al. 2014).

Central America has been identified as a 'hot spot' for climate change (Giorgi 2006) and the Pacific Coast of Nicaragua will increase its vulnerability to extreme climatic events in the future (Hidalgo et al. 2013). Climate change predictions for Central America and the Caribbean show a 2°C to 3°C increase in temperature by mid-21st century and decrease in precipitation (Stocker et al. 2013). Regional projections estimate decreases in precipitation between 7% and 18% for Nicaragua (SICA 2010). Global climate change projections indicate a reduction between 10% and 20% in annual river runoff and water availability for Nicaragua by mid-century (Kundzewicz et al. 2008). Vammen et al. (2012) summarize results from local climate change projection which project a decrease in precipitation for Nicaragua, increase in the frequency of extreme events such as droughts and floodings and increase in mean sea level (Vammen et al. 2012). The last IPCC report (Stocker et al. 2013) establishes that 'it is very likely that for over 95% of the world oceans, regional relative sea level rise will be positive'. The increase in sea level will affect wetlands in Central America. Tropical wetlands sequestrate carbon and help mitigate the effects of climate change (Mitsch and Hernandez 2013). If sea level rise is not accompanied by equivalent vertical accretion of sediments, coastal marshes and mangroves might disappear due to inundation and erosion. Consequently, the socio-economic and environmental services of this ecosystems will be lost (Mitsch and Hernandez 2013).

In Central America, hydrological research and water resources assessment is hindered by the lack of human and financial resources. The general approach for catchment baseline studies emphasizes on soil and forest management and little resources are spent on analysis of water resources system (Faustino and García 2001, Faustino et al. 2007). However, a deeper scientific understanding of water resources and their variability in space and time is needed to create more effective sustainable river basin management plans (Benegas et al. 2008). Still, this understanding is basically missing in Central America.

Basic water resources assessments are normally done in short time periods, often not even covering one hydrologic year. Certain catchment characteristics, such as land use, socio-economic factors and point source contamination sites are more deeply investigated than hydrological processes. In most cases, basins are ungauged or have discontinuous hydrometeorologic records. Few examples of long-term hydrological studies can be found in the region: Plan Trifinio, which covers the transboundary Lempa catchment shared by Guatemala, Honduras and El Salvador and has started in 1988 (Llort and Montufar 2002); PROCUENCA, which covered the transboundary Rio San Juan catchment between Nicaragua and Costa Rica and started in 1995 through 2004; the natural resources management plan for El Cajon catchment in Honduras, which spanned from 1989 to 1991 (OEA 1992); and PIMCHAS in Nicaragua (Orozco et al. 2008). Although these works are valuable sources of data and provided baseline studies to develop integrated water resources management plans, they still lack an in–depth hydrological processes analysis and they are limited to an assessment of water availability and water quality.

Official water resources management plans have not been published yet in Nicaragua. Several initiatives from non-governmental organizations have designed plans but mostly for small catchments. However, these plans need to be approved by the corresponding authorities in order to be enforced. Additionally, catchment hydrology studies in Nicaragua are rarely found in the literature; two examples are Mendoza and Barmen (2006) and Parello et al. (2008), although Parello *et al.* (2008) focused on crater lakes. More studies are found regarding aquifer investigations using mainly numerical modeling and geophysical techniques (Mendoza et al. 2008, Calderon and Bentley 2007, Corriols and Dahlin 2008, Sequeira Gómez and Escolero Fuentes 2010).

All these challenges stress the need to develop and improve the scientific quality of hydrological research in Nicaragua. This has to begin with the improvement of the hydrometerological network and strengthening of local scientific capacities. Long-term hydrological studies are necessary, not only for scientific reasons, but also for the socio-economic and environmental benefits derived from proper water resources management based on sound hydrological knowledge.

1.4. Study area

1.4.1. Geology and geomorphology

The Central American isthmus encompasses highly variable tectonic, lithologic and climatic conditions which resulted in various geomorphologic regions (Marshall 2007). Nicaragua is located in the southern part of the Chortis block within the Caribbean plate. The Cocos plate is subducted beneath the Caribbean plate and the associated Middle America Trench is about 100 km west of the Pacific Coast of Nicaragua (Fig. 1.2). The Middle America subduction zone between the Cocos and Caribbean plate generates earthquakes, volcanism and plate deformation (Elming and Rasmussen 1997, Marshall 2007).

Four main geomorphologic provinces have been distinguished in Nicaragua (McBirney and Williams 1965): the Pacific Coastal Plain, the Nicaraguan Depression, the Interior Highlands and the Atlantic Coastal Plain (Fig. 1.3). The Interior Highlands are composed of Tertiary to upper Cretaceous volcanic rocks, superimposed on methamorphosed Paleozoic and Mezosoic rock formations. The narrow strip of the Pacific Coastal Plain is formed by volcanic and sedimentary rocks from Pliocene to upper Cretaceous. The Nicaraguan Depression is defined as a half graben, it separates the Pacific Coastal Plain from the Interior Highlands. The depression is filled by Quaternary to Mio-Pliocene volcanism products and Quaternary sediments (Elming and Rasmussen 1997, Elming et al. 2001).

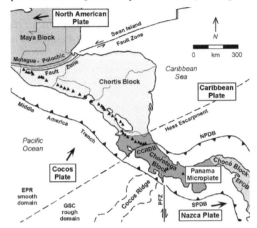

Figure 1.2 Tectonic map of Central America showing the Middle America Trench subduction zone, the Cocos and Caribbean plates and the Central American volcanic range. Solid lines are active plate boundaries, dashed lines are major bathymetric features. EPR: East Pacific Ridge; GSC: Galapagos Spreading Center; PFZ: Panama Fracture Zone; CCRBD: Central Costa Rica deformed belt; NPDB: North Panama deformed belt; SPDB: South Panama deformed belt and EPDB: East Panama deformed belt (Marshall, 2007)

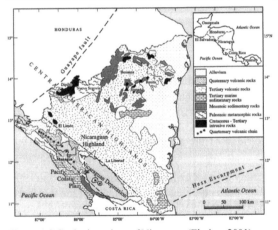

Figure 1.3 Geologic regions of Nicaragua (Elming, 2001)

The study area is located in the geomorphologic region of the Pacific Coastal Plain. This region is formed by a volcano-sedimentary continuous succession, divided into five formations based on lithostratigraphy, petrography and paleontology (Kumpulainen 1995, Elming and Rasmussen 1997). The formations include Rivas, Brito, Masachapa, El Fraile and El Salto. Lithology is predominantly sedimentary; mainly composed of tuffaceous shale, siltstone, greywacke, sandstone and limestone (McBirney and Williams 1965). These formations form a west dipping sequence and are exposed parallel to the Pacific Coast.

The Ostional catchment (Fig. 1.4) is located in a region developed from the deposition of sediments during sea regression and transgression. There are two geologic formations in

9

the area: the Upper Cretaceous Rivas and the Eocene Brito formations (Fig. 1.5a). The Rivas formation of about 2,700 m thickness is composed of drab, tuffaceous shale, sandstone, arkose and greywacke (Swain 1966). The Brito formation is also about 2,700 m thick, overlies concordantly the older Rivas formation. The Brito formation is comprised of volcanic breccias, tuffs, shales and limestones containing orbitoid foraminifera and is brighter in color and more resistant to erosion than the underlying Rivas (Krasny and Hecht 1998, Swain 1966). Volcanic events also influenced its lithology, producing andesitic-basaltic lava flows (Kuang 1971).

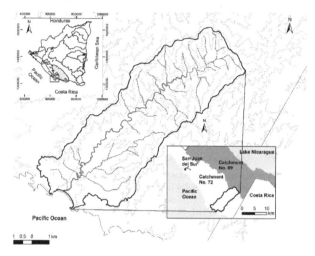

Figure 1.4 Nicaragua's official map of catchment areas (INETER) and location of study catchment. Resolution of elevation contours is 100 m

Tectonic activity in this subduction zone caused compression forces which formed the Rivas Anticline, oriented from the NW to the SE (Fig. 1.5b). The oldest Rivas formation is exposed at the core of this large anticline along the southwestern part of Lake Nicaragua (Kumpulainen 1995, Swain 1966). The Brito Formation is exposed west and north of the Rivas anticline (Swain 1966). The compressional forces also caused faulting and fracturing, generating a fracture system parallel to the ridge of the anticline (Krasny and Hecht 1998, Swain 1966).

The Brito Formation dominates the study area. According to Krasny and Hecht (1998) both Rivas and Brito may have high secondary porosity. However, the depth of the weathered and fractured rock is unknown. Additionally, small valleys contain shallow and narrow aquifers formed by alluvial and colluvial deposits (Fenzl 1989). Transmissivity values reported for the Brito Formation range between 6.3 m^2 d^{-1} and 240 m^2 d^{-1} (Krasny and Hecht 1998). Higher transmissivities (170 m^2 d^{-1} – 603 m^2 d^{-1}) can be expected from quaternary deposits overlying the Brito formation along the rivers, based on data from nearby San Juan del Sur (Krasny and Hecht 1998).

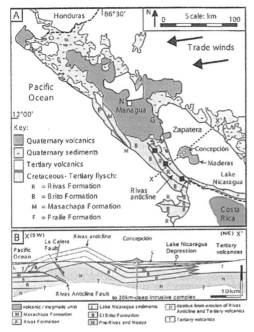

Figure 1.5 a Schematic geologic map of southwestern Nicaragua and **b** Cross section reconstruction of the Rivas Anticline (Borgia and van Wyk de Vries 2003)

1.4.2. Climate

Central America shows a large variability of rainfall distribution, which is largely caused by the orographic effects at the central volcanic cordillera and the position of the coastal line in relation to the trade winds (Hastenrath 1967). Despite the improvements of global weather data supported by satellite observations, *in situ* measurements are still required. Nonetheless, in Central America, long term observations of precipitation have been restricted by complex topography, scarce economic resources and past political conflicts (Durán-Quesada et al. 2012).

The climate of the region is grouped into Koppen AW classification 'tropical dry and wet'. The rainy season starts in May and ends in November. September and October are usually the rainiest months at the Pacific Coast. Usually between July and August, there is a 3 to 5 week period known as midsummer drought (MSD) (Magaña et al. 1999). Precipitation over the Pacific region has a strong sea influence and midday heating is the most probable cause for convective precipitation.

Historical climatic data for the period 1965–2007, from a station located 40 km NW from the study area (Rivas) at an elevation of 70 m asl, registers a mean temperature of 27.1°C, with a minimum of 24.2°C registered in February and a maximum of 30.6°C registered in May. Wind direction is predominantly towards the East and average wind velocity is 5 m s^{-1}. Mean annual precipitation is 1476 mm year^{-1} and mean pan evaporation is 1976 mm year^{-1}. Rainy season starts in May and ends in November, being the rainiest months September and October with 215 mm month^{-1} and 267 mm month^{-1}, respectively.

1.4.3. Catchment structural characteristics

Krasny and Hecht (1998) characterized the hydrologeology and hydrochemistry of the Pacific coast of Nicaragua at a 1:250,000 scale. In this regional study, the area of Ostional is described as a sedimentary formation called Brito. This formation is composed of volcanic breccias, tuffs, shales and limestones. They describe the presence of alluvial deposits in small valleys and secondary porosity in the higher elevation areas. Transmissivity for the Ostional area was reported as 2–33 $m^2 d^{-1}$.

CIRA (2008) performed monthly measurements of river discharge in Ostional using a Pryce AA flow meter with the area–velocity method between August 2007 and May 2008. Highest discharge for the Ostional outlet was reported as 13.4 $m^3 s^{-1}$ in October 2007 during the rainy season; lowest discharge was reported for April and May 2008 as 0.12 $m^3 s^{-1}$. Hydrochemical water types were reported as calcium bicarbonate (CIRA 2008).

Land use (Fig. 1.6) is dominated by forest (52%), agriculture (20%) and pasture (28%) (UNA 2003). Elevation ranges between sea level and 500 m asl in the NE. The upper and middle catchment areas are characterized by narrow V-shaped valleys and steep slopes (20–100%). Population in this area is scarce and dispersed. The lower catchment area is characterized by a 1 km wide alluvial valley and slope below 1%. Land use is a combination of forest on the slopes, agriculture and pasture in the flatter areas.

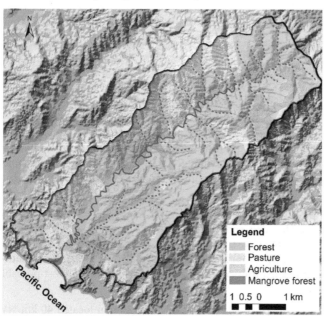

Legend
- Forest
- Pasture
- Agriculture
- Mangrove forest

1 0.5 0   1 km

Figure 1.6 Topography (INETER) and land use (UNA, 2003) of the study catchment

1.4.4. Water resources situation

Population in the catchment is around 1,500 people divided into three communities: Ostional, at the coast; Montecristo in the middle of the catchment; and San Antonio further upstream. Main economic activities are tourism and fishery in Ostional, and agriculture and grazing in Montecristo and San Antonio. Ostional relies on a single well for water supply; however, it suffers from continuous breakdowns which forces people to use groundwater from shallow excavated and unprotected wells. Montecristo and San Antonio rely entirely on these types of wells. No sewage treatment system exists for either community. In the upstream communities only pit latrines are used, gray water from kitchen and showers are disposed directly into the environment. There are about 100 pit latrines in the town of Ostional, but most of them (about 70) are in poor sanitary and structural conditions. From a total of 90 septic tanks, 30 are defective and present leakages (Weeda 2011). Additionally, the town is located at the flood plain of the Ostional River, and according to the National System of Disaster Prevention (SINAPRED) is prone to flooding during high precipitation season (September–October) (SINAPRED 2005).

The South Pacific Coast of Nicaragua has great touristic potential and real estate development is taking off quickly. The investigated Ostional catchment is located within this area and has an extraordinary coastal and rural touristic potential. Nevertheless, increase in tourism and other related developments will create large stress on water resources in this region. The study area is also a typical case of a poorly gauged, tropical forested catchment on the Southwestern Coast of Nicaragua. Catchments in this region exhibit similar hydro-climatic and physiographic conditions (CIRA 2007, CIRA 2008), which makes Ostional a reference catchment for the hydrological research in this region of country.

1.5. Problem statement and objectives

Understanding of the hydrological, socio-economic and legal framework is fundamental for water resources managers since it contributes to the required knowledge for sound decision making, along with stakeholder participation (Falkenmark 2004, Cohen et al. 2006, Kongo et al. 2007, Gooch 2010, Sivapalan et al. 2003).

Hydrological understanding requires mapping the spatio-temporal variability of water resources (Westerberg et al. 2014). Also, it is necessary to understand the connection between catchment structure, hydro-climate and catchment response. This would help to advance hydrologic understanding and to improve predictive capabilities (Wagener et al. 2007, Sivapalan 2006).

Catchments in humid tropics experience highly variable hydrological conditions due to higher energy inputs compared to temperate regions (Wohl et al. 2012). The humid tropical regions are also under strong socio-economic stresses caused by rapid increase in population which leads to fast and sometimes unplanned changes in land use. All of this affects the hydrologic regime. Furthermore, catchments in these regions are usually ungauged, which makes prediction of hydrological changes more difficult and also limits the ability to manage water resources (Sivapalan et al. 2003). Extrapolation of hydrologic knowledge from gauged to ungauged catchments needs intercomparative studies in different hydro-climatic regions to be able to extract similarities and differences (Hrachowitz et al. 2013, Singh et al. 2014).

Central America, and Nicaragua in particular, are examples of these humid tropical regions with enormous challenges to manage water resources. Inadequate hydrometeorological monitoring networks preclude hydrological understanding in most catchments. Actions directed to improve water resources management are usually short-termed and triggered by emergency situations. Although in the case of Nicaragua, a solid legal framework for water resources management is in place, financial and human resources are still insufficient to implement it at its full. Therefore, it is necessary to implement hydrological process studies which help obtain knowledge transferrable to other catchments and be able to establish the links between climate, catchment form and catchment response.

The overall objective of this thesis is to understand the surface and subsurface runoff generation processes in a poorly gauged coastal catchment in Nicaragua under variable humid tropical hydro-climatic conditions. The specific objectives are to:

i.   Identify geomorphological characteristics which control catchment response;

ii.   Investigate hydro-climatic controls on surface and subsurface runoff generation processes at different temporal and spatial scales;

iii.   Study the link between hydrological processes and ecosystem conditions (*i.e.* mangrove forest); and

iv.   Analyze the implications of runoff generation processes for water resources management of the area.

1.6. Thesis outline

The thesis is organized in seven chapters. Chapter 1 provides an overview of water resources management in poorly gauged catchments and hydrological challenges in the humid tropics. It also describes the general water resources situation in Nicaragua and it provides a description of the study area.

Chapters 2 through 6 are based on papers published or accepted (3 and 4) or submitted (2, 5 and 6) to peer-reviewed journals. Chapter 2 investigates groundwater flow systems using a combination of geophysical, hydrochemical and isotopic methods. Chapter 3 examines the hydrological and geomorphological controls on the water balance of the mangrove forest during the dry period. The climatic water balance for the catchment for the period of 2010-2013 is determined in Chapter 4, along with runoff components based on hydrograph separation. Chapter 5 discusses the experience gained from field scale experiments using bacterial DNA as natural hydrological tracers. Chapter 6 looks into seasonal river–aquifer interactions in this area which is dominated by tidal sand ridges. Finally, Chapter 7 provides overall conclusions on the geomorphologic and hydro-climatic controls on catchment response, implications for sustainable water resources management, the link between hydrological processes and ecosystems; and the implications of these results for other catchments in the region. This is followed by recommendations for future research.

Chapter 2

Integrating geophysical, tracer and hydrochemical data to conceptualize groundwater flow systems in a tropical coastal catchment

Abstract

Conceptualization of groundwater flow systems is necessary to improve data collection, numerical modeling, and water resources planning. Geophysical, hydrochemical and isotopic characterization methods were used to investigate the groundwater flow system of a multi-layer fractured sedimentary aquifer along the coastline in Southwestern Nicaragua. A geologic survey was performed along the 46 km² catchment. Geophysical characterization using electrical resistivity tomography (ERT) was applied along a 4.4 km transect parallel to the main river channel to identify fractures and determine aquifer geometry. Additionally, three cross sections in the lower catchment and two in hillslopes of the upper part of the catchment were surveyed using ERT. Stable water isotopes, chloride and silica were analyzed for springs, river, wells and piezometers samples during the dry and wet season of 2012. Indication of moisture recycling was found although the identification of the source areas needs further investigation. The upper-middle catchment area is formed by fractured shale/limestone on top of compact sandstone. The lower catchment area is comprised of an alluvial unit of about 15 m thickness overlaying a fractured shale unit. Two major groundwater flow systems were identified: one deep in the shale unit, recharged in the upper-middle catchment area; and one shallow, flowing in the alluvium unit and recharged locally in the lower catchment area. Recharged precipitation displaces older groundwater along the catchment, in a piston flow mechanism. Geophysical methods in combination with hydrochemical and isotopic tracers provide information over different scales and resolutions, which allow an integrated analysis of groundwater flow systems. This approach provides integrated surface and subsurface information where remoteness, accessibility, and costs prohibit installation of groundwater monitoring networks, which is the case for many catchments in Central America.

Based on: Calderon, H., Flores, Y., Corriols, M., Sequeira, L. and Uhlenbrook, S., in review. Integrating geophysical, tracer and hydrochemical data to conceptualize groundwater flow systems in a tropical coastal catchment. Environmental Earth Sciences.

2.1. Introduction

Groundwater composes about 95% of the Earth's non-frozen freshwater resources. However, despite its importance, global groundwater data collection has received less attention than the more visible surface water resources (Alley et al. 2006). There is still need to better represent groundwater as an integral part of the hydrological cycle. Coupled models of surface water and groundwater are needed to improve simulations of the hydrological continuum (Alley et al. 2006). Conceptualization of groundwater flow systems requires characterization of the geological, hydrological and hydrochemical frameworks (Plummer et al. 2013). This understanding provides information to guide data acquisition strategies, develop numerical models and improve water resources planning (Plummer et al. 2013). For instance, water security evaluation requires a clear inclusion of the prominent role of groundwater storage since aquifers buffer surface water droughts (Foster and MacDonald 2014).

Groundwater flow systems are defined by the distribution of recharge and the boundary conditions controlled by the geologic and topographic setting (Winter 1999). The flow systems can be local, intermediate and regional within the groundwater basin (Tóth 1963). When local groundwater systems are superimposed into the regional system, complex groundwater–surface water exchange patterns can be observed (Winter 1999). Besides the hydroclimatic conditions, geology also influences these exchange processes, defining the spatial distribution of hydraulic conductivity and, therefore, hydraulic connectivity between surface water and groundwater.

The understanding of groundwater flow in fractured media requires the use of many approaches (Berkowitz 2002). Geologic surveying and hydrochemistry has been used to identify groundwater flow systems (Praamsma et al. 2009, Mul et al. 2007, Banks et al. 2009). Hydrochemical and isotopic data are useful to identify recharge sources and groundwater flow systems (Glynn and Plummer 2005). Combined analysis of chloride, electrical conductivity and stable isotopes has been used to differentiate groundwater flow systems in fractured media, e.g. Saha et al. (2013). They found rapid groundwater circulation through fractured zones receiving abundant recharge from precipitation in India. Plummer et al. (2004) used chemical and isotopic data to map groundwater flow systems in porous media in New Mexico. Through the analysis of extensive hydrochemical data they identified and mapped 12 water sources for the basin.

Hydrogeophysical methods can also provide data which might improve the possibilities of process-based modeling and hydrological predictions in basins where hydrometric time series are short or not available (Banks et al. 2009, Rubin and Hubbard 2006, Koch et al. 2009, Francese et al. 2009). Hydrochemistry of spring water offers the chance to assess groundwater in geologically complex catchments. Specially in upland areas where remoteness, accessibility and costs precludes installation of observation wells (Salemi et al. 2013). Moreover, stable isotopic composition of waters has been used to constrain groundwater recharge in fractured aquifers (Abbott et al. 2000). Combined used of hydrometry, tracers and geophysics helps identify runoff generation processes (Wenninger et al. 2008, Uhlenbrook et al. 2008).

Nicaragua's water law mandates the creation of water resources management plans for each one of the catchments in the country, explicitly including assessment and protection of groundwater resources. Development of such plans requires an integrated understanding of both the surface water and groundwater resources of the catchment. However, few hydrogeological studies have been carried out in Nicaragua. MacNeil et al. (2007) used electromagnetic methods to study a caldera in Nicaragua. Calderon and Bentley (2007) used a numerical model to study groundwater flow systems and groundwater extraction scenarios in Northwestern Nicaragua. Mendoza et al. (2008) used electrical resistivity for hydrogeological mapping in a fractured aquifer in a mining area in central Nicaragua to investigate the connection between a polluted river and the aquifer in the Rio Artiguas basin. Sequeira Gómez and Escolero Fuentes (2010) used continuous electrical vertical sounding in combination with geologic survey to study porous and fractured aquifers in central Nicaragua (Malacatoya basin). Significant work using a combined hydrogeological and geophysical approach has also been done in porous aquifers in Northwestern Nicaragua (Corriols and Dahlin 2008, Corriols et al. 2009).

The Ostional catchment is a typical example of a coastal forested catchment of the Southwestern Pacific Coast of Nicaragua. This area is rapidly evolving due to tourist developments. Local communities rely on groundwater for water supply. Nevertheless, as most of catchments in Nicaragua, this is a poorly gauged area where hydrological understanding of surface water and groundwater flow systems is missing. Calderon and Uhlenbrook (2014a), already showed the importance of groundwater contribution for total discharge in the Ostional river. Additionally, Calderon and Uhlenbrook (2014b) presented a model of the river cross section at the catchment outlet, which indicates a strong interaction between the river and the aquifer. Sustainable development of water resources in this area requires the construction of a conceptual model of the groundwater flow system in order to understand the recharge areas and determine appropriate water quality protection measures and the sustainable use of the available water resources.

The objective of this work was to combine non-invasive, integrative methods to identify groundwater flow systems in a remote and geological complex area without a groundwater monitoring network. We aim to develop a conceptual model of the groundwater flow system in a fractured sedimentary aquifer by integrating: i) geophysical surveys with ii) analysis of spatial and temporal variations in surface water and groundwater hydrochemistry and isotopic content.

2.2. Study area

Catchment areas in Central America are identified through a numbering system. Catchment areas on the Atlantic Coast are assigned odd numbers, starting with 1 in Guatemala up to 121 in Panama. Catchment areas in the Pacific Coast are assigned even numbers, starting with 2 in Guatemala up to 164 in Panama (PNUD and OMM 1972). Nicaragua is officially divided into 21 major catchment areas. The study area is located in the catchment No. 72, in the geologic province of the Pacific Coast; west of the Nicaraguan Depression, a NW–SE trending graben where the Managua and Nicaragua Lakes are located (Elming et al. 2001). The depression, described as a half graben, was caused by the subduction of the Cocos Plate below the Caribbean plate with northeastward motion, producing also a volcanic chain in the west side of the depression (McBirney and Williams 1965).

The Pacific Coast is formed by a volcano-sedimentary continuous succession, divided into five formations based on lithostratigraphy, petrography and paleontology (Kumpulainen 1995, Elming and Rasmussen 1997). The formations include Rivas, Brito, Masachapa, El Fraile and El Salto. Lithology in the area is predominantly sedimentary; mainly composed of tuffaceous shale, siltstone, greywacke, sandstone and limestone (McBirney and Williams 1965). These formations form a west dipping sequence and are exposed parallel to the Pacific Coast.

The Ostional catchment (Fig. 2.1) is located in the geologic sub-province known as Rivas–Tamarindo. This region developed from the deposition of sediments during sea regression and transgression. Formations present in the area are the Upper Cretaceous Rivas and the Eocene Brito formations. The Rivas formation of about 2,700 m thickness is composed of drab, tuffaceous shale, sandstone, arkose and greywacke (Swain 1966). The Brito formation is also about 2,700 m thick, overlies concordantly the older Rivas formation. The Brito formation is comprised of volcanic breccias, tuffs, shales and limestones containing orbitoid foraminifera and is brighter in color and more resistant to erosion than the underlying Rivas (Krasny and Hecht 1998, Swain 1966). Volcanic events also influenced its lithology, producing andesitic-basaltic lava flows (Kuang 1971).

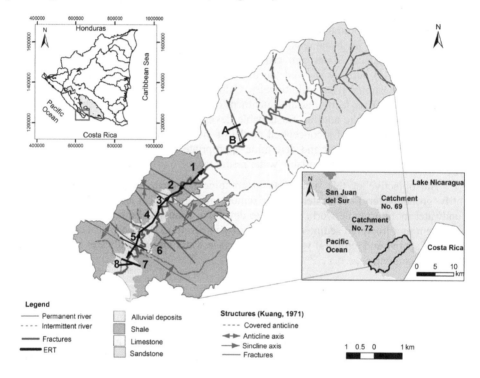

Figure 2.1 Geological features and lithology of the Ostional catchment with location of ERT profiles and water sampling sites. ERT cross sections A and B are located along spring S11. Structures mapped by Kuang (1971) and this study

Tectonic activity in this subduction zone at the contact of the Caribbean and Cocos Plates, caused compression forces which formed the Rivas Anticline, oriented from the NW to the SE. The oldest Rivas formation is exposed at the core of this large anticline along the southwestern part of Lake Nicaragua (Kumpulainen 1995, Swain 1966). The Brito Formation is exposed west and north of the Rivas anticline (Swain 1966). The compressional forces also caused faulting and fracturing, generating a fracture system parallel to the ridge of the anticline (Krasny and Hecht 1998, Swain 1966).

The Brito Formation dominates the study area. According to Krasny and Hecht (1998) both Rivas and Brito may have high secondary permeability. However, the depth of the weathered and fractured rock is unknown. Additionally, the small valleys located towards the south contain shallow and narrow aquifers formed by alluvial and colluvial deposits (Fenzl

1989). Transmissivity values reported for the Brito Formation range between 6.3 $m^2 d^{-1}$ and 240 $m^2 d^{-1}$ (Krasny and Hecht 1998). Higher transmissivities (170 $m^2 d^{-1}$ – 603 $m^2 d^{-1}$) can be expected from quaternary deposits overlying the Brito formation along the rivers, based on data from nearby San Juan del Sur (Krasny and Hecht 1998).

The study area is also a typical case of a poorly gauged, tropical forested catchment on the Southwestern Coast of Nicaragua. Catchments in this region exhibit similar hydro-climatic and physiographic conditions (CIRA 2007, CIRA 2008), which makes Ostional a reference catchment for the hydrological research in this region of country. The upper and middle catchment areas are characterized by narrow V-shaped valleys, steep slopes (20–100%) and no alluvium. Population in this area is scarce and dispersed and the predominant land use is forest. The lower catchment area is characterized by a 1 km wide alluvial valley of an estimated thickness of 15 m (Calderon et al. 2014) and slope below 1%. Land use is a combination of forest on the hillslopes, agriculture and pasture. Population (1,500 inhabitants) is concentrated around a small town called Ostional.

2.3. Materials and methods

2.3.1. Geological characterization

Data analysis for geological interpretation included aerial photos (1:40,000), a Digital Elevation Model (DEM) from Shuttle Radar Topography Mission (SRTM) with a 30 x 30 m^2 grid, and geologic and topographic maps at 1:50,000 scale. A geologic survey was carried out between 2011 and 2012. Local geology was mapped at a scale of 1:10,000. Representative rock samples were collected along the catchment for thin section analysis in the lab.

2.3.2. Electrical Resistivity Tomography surveys

Five Electrical Resistivity Tomography (ERT) profiles (ERT1 to ERT5) were done at a 4,500 m long transect from NE to SW (Fig. 2.1). The ABEM Lund Imaging System (Dahlin 1996) was used with a Schlumberger array with a spacing of 5 m for a penetration depth of 110 m; and 10 m for a penetration depth of 60 m. The system is based on a combination of lateral electrical profiling and vertical electric sounding. Components of the system are: a computer, Terrameter SAS4000 with four input channels (measuring unit), electrode selector unit, multicore electrode cables, and steel electrodes. Data was collected through a roll along technique where cables are moved upward or downward along a succession of stations according to a defined array.

Three cross sections were surveyed at the lower part of the catchment from W to E direction (ERTs 6, 7 and 8). Lengths of these sections were 1,000 m, 800 m and 300 m (Fig. 2.1). Survey penetration depths are around 60 m, except profile ERT 6 with 10 m spacing, which reached up to 110 m depth. Two ERT cross sections were performed across spring S11 in the upper part of the catchment (ERT A and ERT B in Fig. 2.1). Length of these cross sections was 150 m and maximum penetration depth was 30 m.

The data inversion software used was RES2DINV (Loke and Barker 2004), which generates a 2D model of subsurface resistivity. The 2D resistivity data was interpreted using the robust (L_1-norm) inversion method (Loke et al. 2003). This method minimizes the absolute differences between measured and calculated apparent resistivity.

The resistivity model fitness was evaluated by the mean residuals value which compares the resistivity values calculated by the model and the measured apparent resistivity.

19

ERIGRAPH was used for graphical presentation of 2D resistivity imaging. RockWorks was used to produce 3D panels from the longitudinal profiles and cross sections.

2.3.3. Lithologic sampling

Ten piezometers were drilled at the lower part of the catchment across the river channel at the location of ERT 8 (Fig. 2.1). Piezometer depths increase away from the river on both East and West sides from 2 m to 25 m. Split spoon samples were recovered every 0.7 m for the first 15 m. After this depth, samples were recovered from the overflowing injection water from the rotary drilling. Drilling samples were analyzed at a macroscopic and microscopic level in the lab. The samples were used to develop a stratigraphic column for each piezometer and a stratigraphic cross section. The stratigraphic data was used to interpret ERT profiles. Rock samples collected along the catchment were used to prepare thin sections.

2.3.4. Hydrochemical and stable isotope sampling

Water samples were collected throughout the catchment from 5 river locations, 12 springs, 14 wells (integrated depth samples) and 6 piezometers; during the dry (April and December) and the rainy (June and October) season in 2012. Locations are shown in Figure 2.2. Electrical conductivity (EC), pH and alkalinity (HCO_3^-) were measured *in situ*. Samples were immediately filtered through 0.45 μm glass fiber filters. The samples used for cations (25 mL) analyses were acidified with concentrated H_2NO_3 to prevent precipitation reactions and cation attachment to the surface of the sample bottle. The samples for anions (50 mL) were stored in non-acidified bottles.

Bulk precipitation samples were collected weekly during the hydrologic year 2010–2011 for isotopic analyses of δ^2H and $\delta^{18}O$. The Local Water Meteoric Line (LWML) was constructed from these data. Modeled isotopic data from www.waterisopes.org (Bowen et al. 2005) for our study region were also included in the LWML for comparison.

Deuterium excess (d-excess) was calculated for spring water samples according to Equation 2.1 (Dansgaard 1964).

$$d = \delta^2H - 8\delta^{18}O \qquad (2.1)$$

Chloride was measured using an Ion Chromatography System (IC: Dionex ICS 1000). The calculated ion balance error was between -10% and +10%. Silica was analyzed using the molybdosilicate method (APHA 1998) in a spectrophotometer . The detection limit for the method is 0 to 1.71 mg SiO_2/50 ml). δ^2H and $\delta^{18}O$ were measured in an LGR Liquid Water Isotope Analyzer (LWIA). Analytical error for δ^2H amounts to ±1‰ and for $\delta^{18}O$ to ±0.2‰. All the analyses were performed at the hydrochemical laboratory at UNESCO–IHE in the Netherlands.

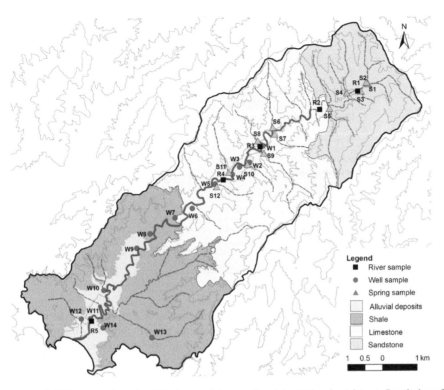

Figure 2.2 Water sampling sites, lithology and topography of the Ostional catchment. Resolution of elevation contours is 100 m

2.4. Results

2.4.1. Stratigraphy

The upper catchment is characterized by rugged topography with numerous steep hills and narrow valleys. Slope varies between 20–100%. The hills are composed of shale and limestone of approximately 30–50 m height with high fracture density (Appendix 2.1). Calcareous sandstone is found at the base of these hills (Appendix 2.2). The predominant strike direction is SW. The sandstone unit dips 10° to 30° SW. The shale unit dips 18° and 35° towards the SW.

In the lower catchment area, only the shale unit is exposed. Outcrops are found near the coast showing high fracture density (Appendix 2.1). On top of the shale, a 15 m thick alluvial deposit was found. The alluvium is composed of fine to coarse sand, gravel, and pebbles. A thin (< 3 m) discontinuous clay layer was observed near the river banks on top of the alluvium. Description of thin sections from rock units are described in Table 2.1.

Table 2.1 Microscopic description of rock units within the study area

Unit	Grain size (mm)	Minerals	Lithic fragments	Matrix	Fossils
Sandstone	0.2-0.8	50%	15%	10%	25%
		quartz, plagioclase, amphibole, pyroxene, biotite	Volcanic igneous	Limestone carbonatic cement	*Foraminifera* (Globigerina) and gastropods (Turritela)
Limestone	<0.004	10%	8%	80%	2%
		plagioclase, potassium feldspar, quartz	Volcanic igneous	Micritic Carbonatic cement	Foraminifera (Globigerina)
Shale	1/16	10%	10%	20%	60%
		quartz, plagioclase, potassium feldspar, biotite	Volcanic igneous	Clay carbonatic cement	*Foraminifera* (Globigerina, Nummulites and Fusulina)

2.4.2. Geology and surface water flow system

Main fractures were identified in catchment area (Fig. 2.1). Other geologic features mapped by Kuang (1971) are also shown. The most relevant features are a longitudinal fracture along the SE water divide, a transversal fracture in the middle catchment area and the Ostional anticline, with its associated fractures. The main strike direction of the fractures is NW-SE.

The compressional forces in the study area cause deformation and fracture of the shale and limestone units with high density of perpendicular joints. The sandstone unit is more compact and less permeable. Intermittent springs originate at the contact between the more permeable shale/limestone and the massive sandstone. Travertine deposits are commonly found at the headwaters of these springs. In general, the springs are located in narrow V–shaped valleys with steep slopes and thin (1–3 m) alluvial deposits at the base of the hillslopes.

The surface drainage network is rectangular, whereby the tributary streams join the main river at an almost right angle. This pattern is created by the presence of fractures as observed in Figure 2.1. In the upper catchment area the main river is fed by springs during the dry season. River flow is intermittent in this area. Water seeps out at the base of shale and limestone walls where they intersect the sandstone streambed. Water seeps into small alluvial deposits and reappears further downstream where the alluvium becomes too thin.

In the lower part of the catchment, a discontinuous clay layer of about 3 m thickness was found on top of alluvial deposits of about 15 m thickness was confirmed through drilling (Calderon et al. 2014). Discharge in the main river channel in this area is permanent, with lowest value of 0.2 m^3 s^{-1} during dry season and a maximum of 5 m^3 s^{-1} during the rainy season of 2012.

2.4.3. Geophysics

ERT profiles are presented in Figure 2.3. Resistivity values range between 3 Ωm and 180 Ωm (Table 2.2). In most profiles a 10–15 m thick layer of lower resistivity is identified (3–16 Ωm). ERT 1, located most upstream, outside the alluvial deposits unit (Fig. 2.1) shows more resistive values on the top, except at the interception with the river, near the center of the profile (Fig. 2.3). ERT 2 also shows high resistivity values of the top until it approaches the hinge of the Ostional syncline (Fig. 2.1). At this point resistivity decreases.

The remaining profiles (ERT 3 to ERT 8) show a low resistivity layer with the afore mentioned values, on top of a more resistive (25–120 Ωm) layer. Blocks of high resistivity materials (>180 Ωm) are observed in all profiles. Discontinuities in these blocks are related to the presence of fractures, filled with less resistive materials. The position of these discontinuities matches the location of the mapped fractures (Fig. 2.1).

ERT 5 ends at the base of a small hill composed of highly fractured shale. Similarly, ERT 7 crossed this hill on the west side of the profile. In both cases the resistivity values are 16–37 Ωm on the top layer. On the other hand, ERT 6 is aligned with an intermittent stream where alluvial deposits are found. Resistivity values on the top layer of ERT 6 towards the West are below 16 Ωm.

Thickness of the more resistive layer was mapped up to 110 m below surface in ERT 6. Results from ERT 8 and piezometer drilling at this location confirmed that this layer is composed of shale at least up to the maximum drilling depth of 25 m.

Slug test performed in the piezometers show little hydraulic conductivity differences between the clay/alluvium and the shale units, with values in the order of 10^{-1} m d^{-1} to 10 m d^{-1} (Calderon et al. 2014). Additionally, four deep (60 m) wells were found in the lower part of the study area. Two of them are used for water supply, one for the Ostional town and one for a future touristic project. Neither drilling records nor pumping tests were available for these wells. However, based on their depths and interviews of the operators, they seem to be located in the shale unit and do not suffer from a drastic drawdown during the dry season.

Table 2.2 Resistivity ranges according to hydrogeologic units

Hydrogeologic unit	Resistivity range (Ωm)
Clay/Saturated alluvium	< 4.5
Alluvium	< 16
Fractured shale/limestone	25 –120
Sandstone	> 180

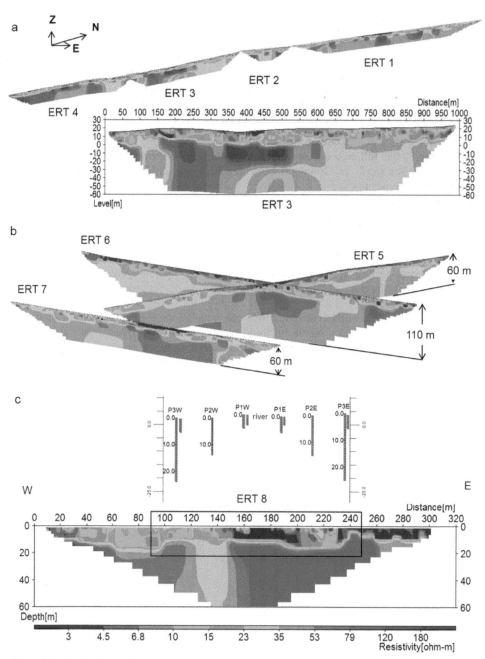

Figure 2.3 a ERT longitudinal profiles ERT1-ERT4. Details of ERT 3 are shown to provide vertical and longitudinal scale; **b** ERT cross sections ERT5-ERT7; **c** ERT 8 at piezometer cross section and piezometer set up. ERT locations are shown in Figure 2.1. Mean residual error varies between 5.7% and 10.5%

2.4.3.1.   Hillslope geophysics

The selected stream (S11) is approximately 2 km long and runs parallel to a fracture. The stream is located in the limestone unit according to the geologic map (Fig. 2.2). Outcrops of highly weathered limestone of approximately 10 m thickness were found upstream (Fig. 2.4a), near ERT A. The stream banks gradually evolve into highly fractured and folded limestone downstream near ERT B (Fig. 2.4b). Stream bed is made of sandstone with some pockets of alluvial deposits (Fig. 2.4c). Thick travertine deposits are observed along the stream (Appendix 2.3).

Figure 2.4 Stratigraphy along Spring S11. **a** Weathered limestone; **b** Fractured and folded limestone at the streambanks; and **c** Sandstone streambed and subsurface water discharge seeping into a small alluvial pocket at the base of the sandstone

ERT A (Fig. 2.5) shows a discontinuity in the high resistivity block, probably sandstone, at a distance of 80 m. The discontinuity is associated with the fracture aligned to the stream (Fig. 2.1). Groundwater seeps out between 90 m–95 m, where low resistivity (7–15 Ωm) materials are observed. The SW side of the profile is located on a steep slope. High resistivity values are observed in the top layer, which corresponds to limestone. A low resistivity material is observed between the limestone layer and the high resistivity sandstone. The same pattern is observed on the NE side of the profile. Between 120 m and 140 m we observe a lower resistivity lense beneath a higher resistivity material.

ERT B (Fig. 2.5) runs parallel to the main river channel. Between 60 and 90 m there is a discontinuity in the highest resistivity material, associated to the fracture along the stream (S11). The discontinuity is filled with more conductive materials. The top layer is composed of weathered limestone.

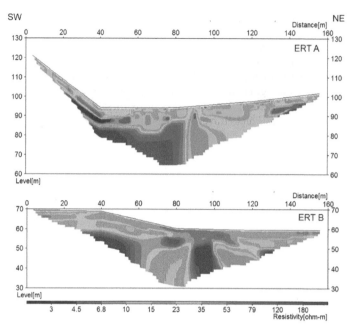

Figure 2.5 ERT profiles A and B at the spring S11, locations are shown in Figure 2.1

2.4.4. Hydrochemical and isotopic characterization of flow systems

Water temperature ranged between 23°C and 31°C, alkalinity values were between 116 mg L^{-1} and 419 mg L^{-1} HCO$^-_3$ and pH values were between 6.9 and 8.3 for all samples during the study period. Areal weighted monthly precipitation for the catchment based on three rain gauges (Calderon and Uhlenbrook 2014a) during the sampling period is shown in Figure 2.6.

EC was consistently higher at the most downstream location (R5); with values between 550 µS cm^{-1} and 612 µS cm^{-1} (Fig. 2.6). Silica values are similar for all locations (Fig. 2.6), below 0.8 mg L^{-1}, except for R1 in October, which reached 1.06 mg L^{-1}. Nevertheless, silica shows a temporal pattern. It decreased around 50% from April to June, but it increased in October during the peak of the rainy season. Chloride concentrations show both temporal and spatial patterns. Location R5 shows always the highest concentrations, but the same temporal pattern as other locations. The temporal behavior shows highest values in April above 10 mg L^{-1}, which decrease in June and October to 6–9 mg L^{-1}. Concentrations increase in December after the end of the rainy season to values between 8 mg L^{-1} and 9.8 mg L^{-1}.

Spring water samples were named S1 to S12, S1 being the most upstream location and S12 the most downstream spring (see Fig. 2.2). EC values were between 335 µS cm^{-1} and 588 µS cm^{-1}. Lowest value was reported for S5 in June. Highest values correspond to S10 and S11 in June and December. The remaining springs vary between 400 µS cm^{-1} and 500 µS cm^{-1}. Silica shows a clear temporal pattern: highest values were observed in April, between 0.8 mg L^{-1} and 0.9 mg L^{-1}, with the exception of S9 which was consistently low (0.3 mg L^{-1} to 0.5 mg L^{-1}). Silica concentrations decreased about 50% in June, and increased in October and December to almost the original values of the dry season (April). Chloride concentrations were between 5.5 mg L^{-1} and 42 mg L^{-1}. Highest concentrations occurred at the most upstream springs (S1 to S4) in December. The remaining springs show higher values in April (dry season), a decrease in June and increase in October in December. However, concentrations in December were still below the April values.

Silica shows spatial and temporal patterns for well water samples. For samples W1 to W10, concentrations varied between 0.35 mg L^{-1} and 0.83 mg L^{-1} with highest values reported for December. For samples W11 to W14, concentrations varied between 0.5 mg L^{-1} and 1.87 mg L^{-1}, with highest values reported either in April or October. W1 to W10 are located in the upper–middle catchment area. W11 to W14 are located in the lower catchment area, around the Ostional town. A spatial pattern was also identified for chloride. For most cases in samples W1 to W10, concentrations were between 10 mg L^{-1} and 14.5 mg L^{-1}. Exceptions are W1, W4 and W10. Samples W11 to W14 showed higher concentrations, between 15 mg L^{-1} and 304 mg L^{-1}. EC values varied between 377 µS cm^{-1} and 1584 µS cm^{-1}. Samples W1 to W10 showed values between 377 µS cm^{-1} and 570 µS cm^{-1}, except for June when they increased to a range between 500 µS cm^{-1} and 1030 µS cm^{-1}. Samples W10 to W14 showed values between 550 µS cm^{-1} and 820 µS cm^{-1}. W14 was an exception with the highest values reported in October and December 1540 µS cm^{-1} and 1584 µS cm^{-1}, respectively.

EC on piezometers screened on the shale unit (P3) show an increase of EC from April to December (Fig. 2.7), from 457 µS cm^{-1} to 1073 µS cm^{-1}. An extreme value was observed in P3W in June (5740 µS cm^{-1}). Piezometers screened in the alluvium show a decrease in EC from April to December. Values on the East side of the river are between 500 µS cm^{-1} and 762 µS cm^{-1}. Values on the West side of the river are higher, between 568 µS cm^{-1} and 4680 µS cm^{-1}. Chloride in samples from the piezometers screened in the shale unit (P3) was relatively high during the dry season: 18.7 mg L^{-1} on the East and 26 mg L^{-1} on the West side of the river (Fig. 2.7). Chloride concentration increased in these piezometers in October to 237 mg L^{-1} on the East and 100 mg L^{-1} on the West side of the river. Piezometers screened in the alluvium unit (P1 and P2) showed chloride concentrations between 1093 mg L^{-1} and 1675 mg L^{-1} in April; they are not shown in Figure 2.7 because of the scale of the figure. The excessively high concentrations were caused by the influence of river water during the dry period. Concentrations in June and October decreased to a range between 11 mg L^{-1} and 38 mg L^{-1}. Silica in alluvium piezometers was between 0.7 mg L^{-1} and 0.9 mg L^{-1} while in shale piezometers it was 0.3 mg L^{-1} in April. Silica concentrations during the rainy season decreased in the alluvium but increased in the shale unit.

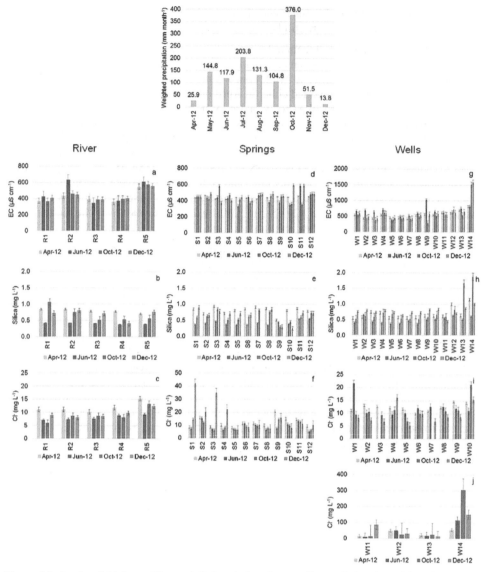

Figure 2.6 Areal weighted monthly precipitation during the sampling period and spatial and temporal hydrochemical variations for different water sources: **a, b, c** River water; **d, e, f** Spring water; **g, h, i, j** Well water

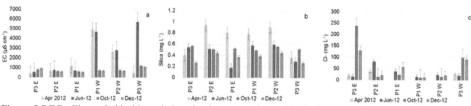

Figure 2.7 EC, silica and chloride variations in piezometers located in the lower catchment area. P1 and P2 are located in the alluvium unit. P3s are located in the shale unit. E and W stand for East and West side of the river, respectively. Piezometer set up is shown in Figure 2.3c. Notice the difference in scales

The Local Meteoric Water Line (LMWL) and the stable water isotope compositions of the different water sources are shown in Fig. 2.8. The LMWL shows ranges from 0.6 ‰ to -90 ‰ for δ^2H and -12.7 ‰ to -1.9 ‰ for $\delta^{18}O$. No altitude nor seasonality effects are observed. Environmental isotope composition of river water samples collected in April shows depletion from sample R1 to R5. Deuterium ranged between -34.5 ‰ to -45.5 ‰.

Spring water samples plot below the LMWL for April, June and December. However, samples collected in October plot parallel above the LMWL (Fig. 2.8). Deuterium excess for spring water samples collected in October (rainy season) is between 11.5 ‰ and 18.3 ‰. Well water samples W1 through W10 show more enriched isotopic contents than samples W11 through W14, both for the dry season (April) and the rainy season (October). Deuterium in samples W1-W10 range between -45 ‰ and -34.7 ‰ in April and between -48.5 ‰ and -45.8 ‰ for October. For samples W11 to W14 the values range between -45.1 ‰ and -41.8 ‰ in April and between -45.5 ‰ and -41.8 ‰ in October.

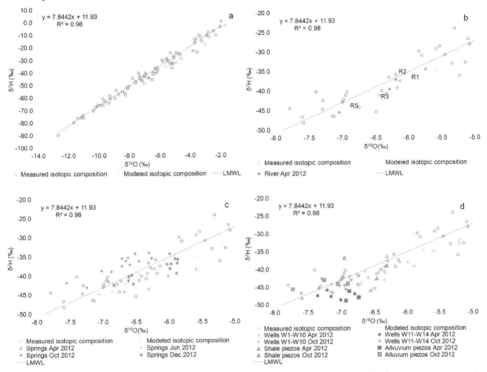

Figure 2.8 a Local Meteoric Water Line (LMWL) and stable isotopic composition of different water sources: **b** river water, **c** spring water and **d** well water

2.5. Discussion

The ERT profiles in the upper catchment area (ERT1 and ERT2) demonstrate the lack of significant alluvial deposits. Resistivity values on the top of these profiles are above 25 Ωm which are associated to the shale unit, based on the drilling samples. Only near the river channel some pockets of low resistivity are observed. As we move downstream from NW to SE, we observe that a low resistivity layer (below 16 Ωm) is developed on the top of the profiles (ERT3). This layer is associated to saturated alluvial deposits on the sides of the main river channel. Underlying the alluvium, a shale unit with intermediate resistivity values (25–120 Ωm) is found. Isolated blocks of high resistivity (above 180 Ωm) are found within the shale, most likely composed of sandstone. Discontinuities in these blocks match the location of the fractures identified from the regional geologic map and the geologic survey. Surface drainage shows a rectangular pattern largely controlled by the fracture network. Springs occur along these structures and intersect the main river channel at nearly right angles as shown in Figure 2.1.

Subsurface water seeps out at the contact between fractured limestone or shale and the compact sandstone. ERT analysis from spring S11 indicates subsurface flow from the steep hillslopes towards the streambed. This is in line with hydrograph separation results from a rainfall–runoff event for this spring, which shows that subsurface stormflow generated because of steep slopes, permeable soils and stratigraphy dominates the hydrograph (Calderon and Uhlenbrook 2014a).

Groundwater flowing to lower catchment area wells shows signs of longer flow paths and residence times, indicated by the high EC values of these samples. However, local anthropogenic influence may be also responsible. EC for river water showed values slightly lower than spring samples. The consistently higher EC at R5 may be caused by a combination of the proximity to the sea and anthropogenic sources, given the vicinity of the town.

Silica concentrations for river water showed a strong dilution after the beginning of the rainy season (June), but concentrations increased again at the peak of the rainy season (October). This increase is explained by the contribution of groundwater to total discharge. Spring water showed higher silica concentrations at the end of the dry season (April). The dilution effect is observed in June. Increased silica concentration in October is the result of discharge of older groundwater.

Groundwater at the upper–middle catchment area showed evidence of local recharge by precipitation. Silica concentrations in well water samples W1-W10 showed lower silica values in April than spring water samples. This indicates that groundwater from these wells has shorter flow paths and residence times than spring water. Diluted silica concentration in April 2012 can be explained by the high precipitation from the previous wet season. Total area weighted precipitation for the catchment in 2011 was 2,103 mm year^{-1}, whereas average historical precipitation for the catchment is 1,476 mm year^{-1} (Calderon and Uhlenbrook 2014a). Further dilution occurred in June 2012 due to local recharge by precipitation. However, concentrations increased in October and reached the highest values in December. This increase can be explained by older groundwater moving downstream the catchment, pushed by the recently recharged precipitation.

The wells in the lower part of the catchment (W11-W14) show silica dilution in June due to local recharge. In October, concentrations increase in wells W13 and W14. Both these wells are associated with fractures and W13 is a deep well (60 m) located in the shale unit. Same behavior is observed in piezometers located in the shale unit (P3E and P3W). High silica concentration is the result of longer flow paths and travel times for these waters. This indicates a deeper groundwater flow system in the shale unit where older groundwater is circulating.

Chloride in river water showed dilution due to precipitation in June and increase in October, except R5 which is located near the ocean and the town. The increased concentrations in December are explained by the contribution of groundwater. Concentrations in the shallow alluvium piezometers show chloride dilution in October caused by local recharge. Average chloride concentration in precipitation samples was 2.5 mg L^{-1}

(Calderon and Uhlenbrook 2014a) compared to 15 mg L^{-1} in groundwater. After the rainy season, chloride concentrations in the alluvium increase again in December. Chloride concentrations in well samples show no clear trend. Probably this is because of mixing of groundwater from the alluvium and the shale unit within the wells. The wells are hand-dug and capture groundwater across both units.

The most upstream springs (S1 to S4) showed a large increase in chloride in December. So did the piezometers located in the shale unit. The relative high chloride concentration in springs and shale piezometers in the dry season indicates that this is characteristic of groundwater in contact with the shale unit. Tweed et al. (2005) also found higher chloride concentrations in groundwater from the marine sedimentary layer of a multilayer fractured aquifer in Australia. The geographical position of our springs at the headwaters of the catchment determines short travel and response times for groundwater (short and steep hillslopes), thus increased Cl⁻ content is observed after the rainy season. This is likely caused by mobilization of high Cl⁻ groundwater through piston flow. Recharged water has to displace all residual groundwater. In constant groundwater density environments, groundwater flow reaches steady state before all residual groundwater has been displaced (Stuyfzand 1999). This means that different chemical compositions and isotopic signatures will be observed within the flow system, as is our case.

Deuterium excess in spring water samples in October is caused by precipitation. The Global Meteoric Water Line (GMWL) has a d-excess value around 10‰ which results from a single stage evaporation from the ocean at a relative humidity of 85% (Clark and Fritz 1997). Evaporation of precipitation intercepted by the canopy (Savenije 1995) may partly be the cause of d-excess in spring waters, since the upper catchment area is mostly covered by forest. Another possible source of moisture is evaporated water from Lake Nicaragua, which is only 10 km NE from the study catchment (see Fig. 2.1). Although sea water evaporation is the principal source of moisture in Central America (Magaña et al. 1999), we cannot discard the influence of the lake on precipitation. Open water bodies and soil moisture have been identified as sources of evaporated water in tropical rainforests (Martinelli et al. 1996). However, infiltration capacity of the soils is high (50 mm h^{-1}) (Calderon and Uhlenbrook 2014a), providing good drainage capacity. Other possible cause for the d-excess is enrichment of precipitation as it falls amount effect), especially during short storms and under low relative humidity conditions (Kendall and McDonnell 1998). Average relative humidity for the study area is 83% for October and daily precipitation was usually above 20 mm day^{-1}, which reduces the chance for the amount effect. These possible causes require further investigation to establish the sources of recycled moisture during the rainy season.

Lachniet and Patterson (2002) used deuterium excess to study moisture recycling in Costa Rica. They found d-excess values up to 21‰. Although their study transect crosses a mountain range with a maximum elevation of 3,600 m asl, they found that elevation is not the dominant control for d-excess, but rather moisture recycling. However, they were not able to differentiate between soil moisture and interception as the sources of re-evaporated water. Rhodes et al. (2006) also found d-excess which increased as the rainy season progressed in a study site in Costa Rica at about 20 km from the coast. One of the causes identified for this deuterium enrichment was water evaporated from a very wet region within the study area. However, they could not identify a specific source of re-evaporated water.

Well water samples in the lower catchment area have a lighter isotopic signature than well water samples from the upper and middle catchment area. The piezometers in the alluvium show a similar isotopic signature than the lower catchment wells. The piezometers in the shale exhibit a similar composition than the upper–middle wells. This difference

31

indicates different groundwater recharge sources. The alluvium unit shows a lighter isotopic composition because receives local recharge from the river, which has undergone evaporation. The shale unit has a heavier isotopic composition indicating upstream recharge, with no influence of evaporated river water.

Recharge from precipitation causes mobilization of older groundwater (longer flow paths and residence times) within the shale unit. This is observed as increase in silica and chloride concentrations in springs and deep piezometers at the peak of the rainy season (October). Conversely, the alluvium is recharged locally by precipitation, and possible by river water; causing dilution of silica and chloride in shallow piezometers.

Groundwater flow systems derived from chemical and isotopic results

Chemical and isotopic results indicate that there are at least two major groundwater flow systems in the catchment: one deeper system in the shale unit, recharged in the upper–middle catchment area; and a shallow system located in the alluvium unit that is recharged locally at the lower catchment area. The upper–middle area is formed mainly by fractured shale on top of compact sandstone with no significant alluvial deposits. The L area is comprised of an alluvial layer of about 15 m thick overlaying a fractured shale unit.

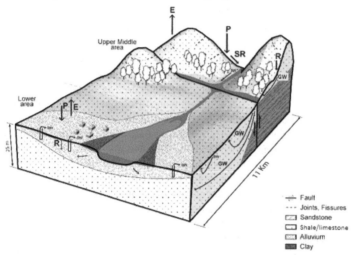

Figure 2.9 Sketch of the conceptual model of groundwater flow systems. P: precipitation, E: evaporation, R: recharge, SR: surface runoff, SW: surface water, and GW: groundwater. Not to scale

2.6. Conclusions

The upper–middle catchment area is characterized by a fractured shale/limestone unit on top of a compact sandstone unit. Alluvial deposits in this area are not significant. The lower catchment area has an alluvial unit on top of a fractured shale unit. Two major groundwater flow systems were identified: one deeper system in the shale unit, recharged in the upper–middle catchment area; and a shallow system located in the alluvium unit that is recharged locally at the lower catchment area. Groundwater in the shale unit is characterized by high chloride concentrations. Recharged precipitation displaces older groundwater along the catchment, in a piston flow mechanism. The surface drainage system is controlled by fractures as shown by the angles of intersection between the springs and the main river. Springs are also controlled by the occurrence of fractures, they occur at the contact between the fractured shale and limestone and the compact sandstone. Indications of moisture recycling were found. However, further investigations are needed to determine the actual sources.

Hydrogeophysical methods provide invaluable insights into areas where no observation or pumping wells are available. Geophysical data supported by hard evidences of rock samples (thin sections) provide a better definition of aquifer stratigraphy. Additional information of stable water isotopes and hydrochemistry provide further insights into the groundwater flow systems.

The combination of non-invasive, integrative methods for groundwater flow system identification is an efficient and practical approach for remote and geological complex areas, where groundwater monitoring network are not financially or logistically feasible. Such is the case of many catchments in Central America, where this approach could be applied.

Appendix 2.1 a Shale wall in the upper catchment, notice high fracture density and **b** Shale outcrop at the coast

Appendix 2.2 Stratigraphic sequence at the spring S11

Appendix 2.3 Travertine deposits at spring S11

Chapter 3

Hydrological and geomorphological controls on the water balance components of a mangrove forest during the dry season

Abstract

Hydrological and geomorphological processes are key to mangrove forest growth and development. However, few mangrove hydrology studies exist in Central America. A 0.2 km^2 mangrove forest on the Pacific coast of Nicaragua was investigated to determine the water balance dynamics during the dry season. The used multi-methods approach combined hydrology, hydrochemistry and geophysics. Precipitation is the main freshwater input. Tidal sand ridges are the key geomorphologic features which allowed an increase in water storage of 351 m^{3}d^{-1} during a 22 day period. Large precipitation events cause breaking of the beach ridges by excess water, suddenly emptying the system. Grey water and pit latrines from the nearby town influence shallow groundwater quality, but also provide nutrients for the mangrove forest. Groundwater chemistry is also affected by calcite dissolution and seawater. Refreshening and salinization processes are controlled by the general groundwater flow direction. Hydraulic and hydrochemical influence of seawater on coastal piezometers seems to be controlled by the elevation of the water table and the tidal amplitude. These conditions control forest subsistence during the dry season, which is essential for the mangrove forest to provide ecological and economic benefits such as protection against flooding, habitat for numerous species, and tourist attraction.

Based on: Calderon, H., Weeda, R. and Uhlenbrook, S. 2014. Hydrological and geomorphological controls on the water balance components of a mangrove forest during the dry season in the Pacific Coast of Nicaragua. Wetlands. 1-13 DOI 10.1007/s13157-014-0534-1.

3.1. Introduction

Wetlands are at the interface between aquatic and terrestrial ecosystems. Wetlands in coastal areas in equatorial regions form mangrove swamps, whereas at higher latitudes they tend to form salt marshes (Keddy 2010). Mangroves can be very broadly defined as an assemblage of plants all adapted to a wet, saline habitat (Feller and Sitnik 1996, Thom 1967). Increased water level caused by climate change is one the many threats to mangrove forest survival worldwide (Gilman et al. 2008, Alongi 2002, Briceño et al. 2013, Polidoro et al. 2010).

Mangrove forests are usually located between mean sea level and the highest spring tide with different mangrove tree species typically occurring in zones parallel to the coast (Alongi 2002). This zonation is attributed to salinity, soil type and chemistry, nutrient content, physiological tolerances, predation and competition (Smith et al. 1992). Thom (1967) established that the most important factors controlling mangrove zonation are substratum and water flow regime. Preservation of dynamic flow regimes is critical to maintain riverine ecosystems (McClain et al. 2013) like mangrove forests. Therefore, knowledge of hydrological and geomorphological processes in mangrove forests are critical for understanding the functioning, as well as conservation, protection, and sustainable management, of these fragile ecosystems.

According to Ellison (2004), wetlands in Central America, including mangrove forests, cover approximately 40,000 km^2 (8% of the total area). However, the exact extent and health of mangroves is not known since ecological studies have been focused on Costa Rica and Panama wetlands (Ellison 2004). Groombridge (1992) estimated between 6,500 km^2 and 12,000 km^2, while Spalding et al. (1997) estimated between 8,000 km^2 and 9,000 km^2, excluding Belize. On the Atlantic coast, mangroves form narrow bands along the coastal plains, whereas on the Pacific coast mangroves extend further inland with extensive forests in major river deltas such as the Estero Real in Nicaragua (Ellison 2004). There are approximately 4,000 km^2 of mangrove forests on the Pacific coast of Central America (Jimenez 1992). Ellison (2004) estimated that there are 2,000 km^2 of mangrove forests on the Atlantic coast of Nicaragua, but only 400 km^2 on the Pacific coast side.

Dry climate mangroves are found on the Pacific Coast where precipitation is below 1,800 mm year^{-1} distributed mainly between May and November as opposed to the more than 2,000 mm year^{-1} at the Atlantic side. Soil salinity increases inland as the influence of tidal flooding is reduced, and salt accumulation is enhanced by evaporation. In this kind of environment, evaporation and freshwater inputs impact growing conditions, species composition, and structural development, thus increasing variability of forest composition from site to site (Jimenez 1992). Freshwater flows also equilibrate nutrient inputs which affects the food web within the mangroves (Briceño et al. 2013). Understanding of sources of freshwater inputs into mangrove ecosystems is vital for their conservation and protection (Gondwe et al. 2010).

Most of the literature on mangroves in Central America is related to biological (Gross et al. 2013, Schumacher 2007), ecological (Rabinowitz 1978, Lovelock et al. 2004, Roth 1992, Pool et al. 1977, Carvalho et al. 1999, Cahoon et al. 2003, Gross et al. 2013, Ellison 2004) and socio-economic characteristics (Fürst et al. 2000). Few studies were found on the relationship of mangrove forests and hydrology. A study in the dry life zone of the Gulf of Honduras investigated mangrove zonation patterns finding a great influence of soil salinity (Castañeda-Moya et al. 2006). Another study in Costa Rica used the water balance to study *Avicennia bicolor* in a dry weather zone, finding a great influence of rainfall and runoff on the structure and function of the forest (Jimenez 1990).

There are few studies on mangroves in Nicaragua (Carvalho et al. 1999, Fürst et al. 2000, Schumacher 2007, Perez et al. 2008). Nevertheless, they address exclusively the ecological and economic value of large mangrove forests. No references were found regarding the hydrological functioning of this type of ecosystem in Nicaragua; even though the water flow regime is recognized as one of the most important factors controlling mangrove forest structure and functioning (Thom 1967). Moreover, the hydrological cycle may influence biogeochemical processes within mangrove forest soils (Sherman et al. 1998).

An example of a small, unstudied mangrove ecosystem was found in the southwestern coast of Nicaragua, south of the city of San Juan del Sur, near the town of Ostional (Fig. 3.1). The estuarine mangrove forest occurs along the floodplain of the river Ostional. This small forest has an approximate area of 0.2 km^2, yet provides valuable economic and ecological services in terms of tourist attraction, protects the area against flooding, and provides habitat for numerous species. This mangrove forest is part of the biological corridor of the Pacific of Nicaragua, providing biological interconnectivity with other ecosystems (CBM 2002). The beaches of the area are nesting sites for an endangered sea turtle species, *Lepidochelys olivacea*.

The objective of this study was to determine the water balance dynamics of a mangrove forest during a dry period. The research focused on (i) determining the sources and fluxes of freshwater which influence mangrove subsistence during the dry season using hydrological, hydrochemical and geophysical methods, and (ii) identifying geomorphological and hydrological controls on the water balance components.

3.2. Study area

3.2.1. General

The town of Ostional is a rural community located on the Pacific Coast of southwestern Nicaragua. It has an estimated population of 1,500 people. There is a water supply system which relies on a single 60 m deep well, which suffers from constant malfunctioning, forcing the population to use shallow hand dug wells. There is no sewage treatment system but there are about 100 pit latrines, with most (70) in poor sanitary and structural condition. Of 90 septic tanks, 30 are defective and leak (Weeda 2011); grey water, generated from laundry, showers and kitchens, is directly discharged to the ground. Agriculture and fishing are the main economic activities in the area. The town is located at the flood plain of the Ostional River, and according to the National System of Disaster Prevention (SINAPRED) is prone to flooding during high precipitation season (September–October) (SINAPRED 2005). Land use is dominated by forest (52%), agriculture (20%) and pasture (28%) (UNA 2003).

3.2.2. Regional geology

The study area is located in the geomorphologic province of the Pacific Coast, within the Rivas–Tamarindo sub-province. This environment originated from deposition of sediments during processes of sea transgression and regression. Volcanic activity also contributed to the formation of this sub-province. Tectonic activities in this subduction zone (the contact of the Caribbean and Cocos Plates), caused compression forces which formed the Rivas Anticline, oriented from the NW to the SE. The Upper Cretaceous Rivas Formation is exposed in the large Rivas anticline along the southwestern part of Lake Nicaragua and consists of about 2,700 m of tuffaceous shale and sandstone. The Rivas Formation is overlain by the 2,750 m thick Paleocene–Eocene Brito Formation. The Brito Formation is exposed west and north of the Rivas anticline and is comprised of volcanic breccias, tuffs, shales, sandstone, and limestone. The compressional forces responsible for the formation of the anticline also caused faulting and fracturing of these two formations creating a fracture system parallel to the ridge of the anticline (Krasny and Hecht 1998, Swain 1966). Only the Brito Formation is exposed in the study area.

3.2.3. Climate, hydrology and hydrogeology

According to the Koppen climate classification, the climate is tropical wet and dry (Aw). The rainy season starts in May and ends in November, with September and October being the wettest months with an average of 215 mm month^{-1} and 267 mm month^{-1}, respectively. During July and August there is generally a 3 to 5 week period of no precipitation known as the midsummer drought (Magaña et al. 1999). Historical climatic data for the period 1965–2007 from the nearest station located in Rivas (40 km NW from Ostional) indicates a mean temperature of 27.1°C. Wind direction is predominantly towards the East with an average wind velocity of 5 m s^{-1}. Mean annual precipitation is 1,476 mm year^{-1} and mean pan evaporation amounts to 1,976 mm year^{-1}.

The upstream catchment area is characterized by exposed fractured shale and limestone and intermittent streams with sandstone streambeds. Downstream, alluvial deposits of unknown thickness are found. River discharge varies between 0.1 m^3 s^{-1} (March) and 13 m^3 s^{-1} (October) (CIRA 2008). The aquifers found in the area are formed by weathering and faulting or fracturing of sedimentary rocks, partially filled by alluvial deposits. Transmissivity values, are less than 500 m^2 d^{-1} (Krasny and Hecht 1998). The thickness of the weathered zone where aquifers might form is not known. Alluvial deposits are found in the valleys. Nonetheless, it is believed that these deposits are only a few meters thick (Krasny and Hecht 1998). According to the regional hydrochemical map, groundwater types found the area of Ostional vary between HCO$_3$–Ca and HCO$_3$–Ca–Mg.

3.3.   Materials and methods

3.3.1. Soil and water sampling

Soil samples were collected at fifteen locations in the study area (Fig. 3.1) at 0.25 m depth intervals up to a maximum depth of 3 m. Grain size analyses were conducted using laser diffraction performed by a FRITSCH laser particle sizer A22. Grain size was used to determine soil type according to the USDA textural classification. Cation exchange capacity (CEC) was estimated according to Appelo and Postma (1993).

Shallow groundwater and surface water samples were collected from the river upstream of the mangrove forest (RU) and at the river outlet during high (RO1) and low tide (RO2). Sea1 and Sea2 samples correspond to high and low tide, respectively. Groundwater was sampled at eight excavated shallow wells (W), one deep (60 m) well (Wsw), and from six piezometers (P; depths 1.8–3 m).   Locations are shown in Figure 3.1. Electrical conductivity (EC), temperature, pH and alkalinity (HCO$_3$$^-$) were determined *in situ*. Samples were filtered through a 0.45 μm glass microfiber filter. Samples for cation analyses were acidified with concentrated H$_2$NO$_3$ to prevent precipitation reactions and cation attachment to the surface of the sample bottle.

Major cations (Mg^{2+}, Ca^{2+}, Mn^{2+}, NH$_4$$^+$, Fe^{2+}) were analyzed using an Inductively Coupled Plasma-Optical Emission Spectrometer (ICP–OES: Perkin Elmer Optima 3000), while Na$^+$ and K$^+$ were measured using a flame Atomic Emission Spectroscope (AES: Perkin

Elmer AAnalyst 200). The major anions (Cl^-, HCO_3^-, SO_4^{2-}, NO_3^{2-}) were measured using an Ion Chromatography System (IC: Dionex ICS 1000). The calculated ion balance error was between -10% and +10%, samples with higher errors were discarded. Analyses were performed at UNESCO-IHE laboratories in Delft, The Netherlands.

Hydrochemistry results were examined using a Piper diagram and the fraction of seawater (f_{sea}) based on Cl^- concentration according to Eq. 3.1 from Appelo and Postma (1993):

$$f_{sea} = \frac{m_{Cl^-,sample} - m_{Cl^-,fresh}}{m_{Cl^-,sea} - m_{Cl^-,fresh}} \qquad (3.1)$$

where m is chloride concentration for the sample, the freshwater end-member and the seawater end-member. Concentration for freshwater was taken from sample Wsw, a deep well located outside the town (Figure 3.1). Concentration for seawater corresponds to sample Sea1.

Correlations between Ca^{2+}, Mg^{2+}, Na^+, K^+, SO_4^{2-}, and Cl^- were also examined. This is a group of ions that can be considered to behave as conservative tracers in the salinization process (Giménez and Morell 1997, Ghiglieri et al. 2012).

3.3.2. Stratigraphy and geophysics

Fifteen shallow piezometers were installed by hand auger drilling within the mangrove forest (Fig. 3.1). A PVC pipe of 0.0254 m diameter was used for the piezometers. Elevations were obtained from a GPS Garmin 62sc, which was calibrated with a local topographic benchmark.

Percussion and rotation drilling was performed to install six deeper piezometers at a cross section of the river (A–A' in Fig. 3.1). Percussion drilling is a technique whereby a drill bit is hammered into the ground, rotary drilling uses a rotating drilling bit. Piezometers depths vary between 5 and 25 m, with depth increasing away from the river on the West and East sides (Fig. 3.2). Percussion drilling was used for the first 10–15 m where relatively soft materials were encountered. Split spoon samples were collected every 0.7 m, providing undisturbed samples. Once harder material was found, the technique was changed to rotation drilling using a tricone with water injection system to recover grinded sediments. Sediment samples were collected from water overflowing the borehole. Samples were analyzed at macroscopic and microscopic levels in the lab to determine sediment composition and derive a stratigraphic cross section.

Electrical resistivity tomography (ERT) was applied at the cross section A–A' to gain further insights into the subsurface structure. The ABEM Lund Imaging System (Dahlin 1996) was used with a Schlumberger array with a spacing of 5 m. This array consists of four collinear electrodes where the two inner electrodes are the receivers (potential) and the outer two are the current (source) electrodes. The potential electrodes remain fixed in position while successive readings for increasing spacing of the source electrodes are taken (ASTM 1970). The ERT profile was 300 m long. The survey depth reached up to 60 m. The data inversion software used was RES2DINV (Loke and Barker 2004), and ERIGRAPH (Dahlin and Linderman 2007) was used for graphical presentation of 2D resistivity imaging.

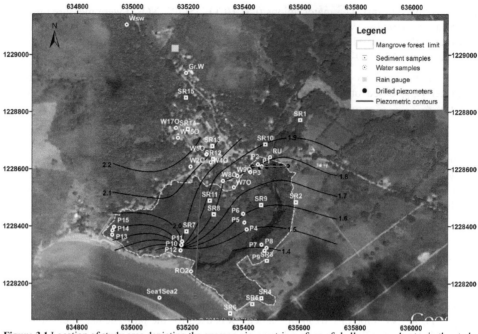

Figure 3.1 Location of study area depicting the average piezometric surface of shallow groundwater in the study period as well as cross sections A-A' and B-B'

3.3.3. Hydrogeologic characterization

Slug tests were performed for nine of the fifteen shallow piezometers. Slug tests are performed by adding or removing water instantaneously from the piezometer (Fetter 2001). Water level changes were measured with a Schlumberger DI501 mini diver (range of 10 m). Data analysis was performed using the method of Bouwer and Rice (1976) for unconfined aquifers and partially penetrating wells. Slug tests were also performed in the deeper piezometers specifically screened at different strata to estimate different hydraulic conductivity (K) values (P3W, P3E and P1W, P1E). Hydraulic conductivity describes the rate at which water can move through a permeable medium (Fetter 2001).

Groundwater levels were monitored continuously in six piezometers (P1, P2, P3, P10, P11, P12) using Schlumberger mini divers. Data was recorded every 15 min from 21 of May through 30 July 2010. River level was also monitored at the A–A' cross section. Barometric compensation of the data was performed using atmospheric pressure data recorded at the same interval and for the same period with a Schlumberger 50013, DI501 barometric diver (range of 1.5 m). Groundwater levels at the remaining piezometers and 21 shallow wells (depth 4.4 m to 11 m) were measured manually six times during the same period. Average groundwater levels from the shallow wells and shallow piezometers were used to derive the piezometric map.

3.3.3.1.   Water Balance Estimation for the Dry Period

Precipitation was measured using a Tenite[TM] rain gauge. A Class A evaporation pan was installed at the same location (Fig. 3.1) to obtain daily evaporation measurements and estimate reference evaporation. The term reference evaporation is defined as the rate of combined evaporation and transpiration from a reference surface (green grass), with no water shortage. This term allows estimation of total evaporation demands of the atmosphere independently of crop type, crop development and management practices. The pan evaporation is related to the reference evaporation by an empirically derived pan coefficient according to Equation 3.2 (Allen et al. 1998):

$$E = K_p\, E_p \tag{3.2}$$

where E is the reference evaporation, E_p is pan evaporation, both in mm day^{-1}; and K_p is the pan coefficient (-), which depends on site specific conditions such as pan type, land cover under the evaporation pan, type of crop surrounding the pan, general wind speed and air humidity conditions. In our case, the coefficient was estimated at 0.7. Measurements of precipitation and evaporation were taken every morning.

The water balance was determined for the period between 21 May to 19 June 2010. Inflow terms included precipitation, groundwater inflow and surface water inflow, while outflow terms included evaporation, groundwater outflow, and surface water outflow. Groundwater extraction was considered negligible since there is only one pumping well, which was not functioning at the time. Groundwater runoff was determined using the Darcy equation (Eq. 3.3):

$$Q = -KA\frac{dh}{dl} \tag{3.3}$$

where Q is runoff (m^3 d^{-1}), K is hydraulic conductivity (m d^{-1}), A is cross sectional area (m^2); dh/dl is the hydraulic gradient (i), given by the hydraulic head difference between two points, dh (m) and the distance between those two points, dl (m).

Surface water inflow and outflow were set to zero for the period of study (dry season) as no flows could be detected. Groundwater storage changes were also considered. The water balance was estimated according to Equation 3.4:

$$P + SW_{in} + GW_{in} = E + SW_{out} + GW_{out} + \frac{\Delta S}{\Delta t} \tag{3.4}$$

where P is precipitation (m^3 d^{-1}), E is evaporation (m^3 d^{-1}), SW is surface water runoff (m^3 d^{-1}), GW is groundwater runoff (m^3 d^{-1}), S is storage (m^3) and t is time (d).

3.4. Results

3.4.1. Coastal geomorphology and mangrove forest characteristics

The Ostional mangrove forest can be classified as riverine, occurring along the Ostional river estuary (Fig. 3.1). The geomorphology of this environment is determined by strong waves and littoral drift, which produces sand ridges along the coast. The mangrove forest is found behind these ridges, and it connects with the ocean trough the Ostional river outlet. During the study period, these ridges prevented direct surface connection between the mangrove and the sea; thus allowing surface water storage as backwater in the estuary (Appendix 3.1).

Zonation patterns of three mangrove species were observed: red mangrove (*Rizophora mangle*) was found on the shore of the estuary with black mangrove (*Avicennia germinans*) further inland. Buttonwood *(Conocarpus erectus)* was found along the coast on higher beach ridges away from the tidal zone.

3.4.2. Soil texture and stratigraphy

All soil samples were classified as clay. Calculated cation exchange capacity (CEC) ranges between 25.95 meq 100 g^{-1} and 37.13 meq 100 g^{-1}.

Sediments collected during drilling on the East side of the river were composed of clay up to depths of 3 m. At location P3E it reached 7 m. Samples reacted vigorously to HCl indicating high calcite content (Appendix 3.2). Clay was in some cases mixed with subangular shale fragments (pebble size) like in P2E, probably the result of fluvial deposition. In the case of P1E and P3E, the clay layer was very homogenous. However, the clay deposits on the West side of the river were also combined with loam and shale fragments. Alluvial deposits were found beneath the clay layer. These were part of a heterogeneous layer composed of medium to fine sand, gravel and pebbles. The alluvium is thicker on the West side of the river, reaching 10 m thickness. Underneath the alluvium, a shale layer was found. Its thickness is unknown since perforations only reached 25 m depth. Although the drilling method did not allow the recovery of unaltered samples of this layer, inspection of an outcrop located 100 m to the north, showed a very high fracture density. Stratigraphic correlation along the cross section is shown in Figure 3.2.

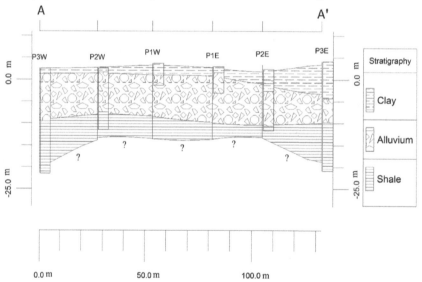

Figure 3.2 Stratigraphic correlation in cross section A-A', vertical exaggeration 2X

Comparison of stratigraphic correlation and the ERT profile (Fig. 3.3) shows a distinctive resistivity for the shale layer relative to the upper alluvium and clay units. Saturated clay sediments showed resistivity values below 4.5 Ωm. Resistivity of dry clay and alluvial deposits range between 6.3 Ωm and 23 Ωm. A higher resistivity was observed on the upper part of P2E, where shale fragments were found. The shale showed resistivity values between 23 Ωm and 79 Ωm. The shale unit is observed as large outcrops at the coast (Appendix 3.3). These observations may indicate that the shale unit pinches out towards the coast.

A slight resistivity difference was observed on the shale unit between west and east piezometers. Resistivity values were slightly lower on the west side, between 43 Ωm and 69 Ωm; whereas resistivity on the east side was above 69 Ωm. Drilling samples from west piezometers showed the same stratigraphy as east piezometers (Fig. 3.3). Since rotation drilling was used for the shale layer, no undisturbed samples could be recovered in order to determine if shale at this site had a greater fracture density. It could also be possible that there is a greater degree of saturation caused by water bearing fractures.

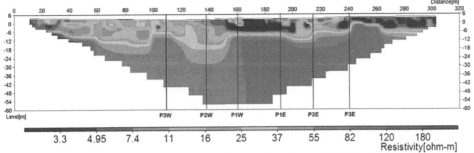

Figure 3.3 ERT profile at the cross section A-A'. Piezometer locations are also included

3.4.3. Hydrogeological system analysis

3.4.3.1.   Saturated hydraulic conductivity

Average hydraulic conductivity (K) values are very similar for all three strata (Table 3.1), within the 10^{-1} m d^{-1} to 10 m d^{-1} range. However, the clay and alluvium strata show lower K and thus, were grouped into one aquifer unit; and the shale into another unit. P3 had a K value of 13.37 m d^{-1} which could be caused by heterogeneities in the clay unit, thus it was not used to estimate the average value.

Since the number of K estimates is limited, we rely on additional evidence to support the distinction between aquifer units. One is the observation of high fracture density of the shale outcrops in the area. The other is the little contrast between the resistivity values of the clay and alluvium; yet the noticeable difference between these layers and the shale.

Table 3.1 Hydraulic conductivity estimates from slug tests, see Figure 3.1 for locations

Stratum	Piezometer	K (m d^{-1})	Average K (m d^{-1})
	P1	0.043	
	P4	1.18	
	P5	0.23	
Clay	P7	0.0006	0.33
	P8	0.001	
	P10	0.88	
	P11	0.05	
	P12	0.29	
Alluvium	P1E	5.60	6.70
	P1W	7.80	
Shale	P3W	12.96	9.07
	P3E	5.18	

3.4.3.2.   Piezometric Surface of Shallow Aquifer and River Stage

General groundwater flow direction is from NW to SE. The piezometric surface for the shallow aquifer is shown in Figure 3.1. Groundwater depth varies between 0.5 m and 5 m. Groundwater levels in piezometers near the coast showed semidiurnal fluctuations due to the effect of sea tides (Fig. 3.4). P10 and P12 show a water level response to tidal effects due to pressure changes in the aquifer (Ferris et al. 1962). However, P11 (most distant from the sea) has a different behavior, responding mostly to precipitation events and showing small tide-induced fluctuations only when the water level is near or below 2 m.

Hydraulic heads dropped because of the natural breakthrough of the river outlet due to increased water level caused by precipitation events. Hydraulic heads in piezometers next to the river (P3 and P1) showed an increase before 24 June 2010 and a sudden decrease on 25 June 2010, like the one observed in P11 (Fig. 3.4). Once the system accumulated sufficient water, the water level overtopped the beach ridges allowing direct surface discharge into the ocean, thus emptying the system and causing a rapid drop in hydraulic heads. Unfortunately, this observation was not recorded since the river mini diver was stolen and surface water data could not be collected during this period.

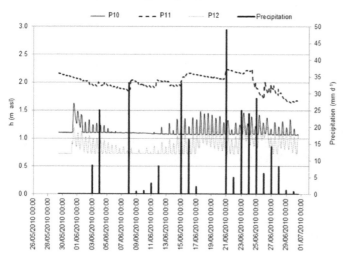

Figure 3.4 Sea tide and precipitation effects in hydraulic heads in piezometers located near the coast. P10 and P12 depict stronger tidal influence due to more proximity to the shoreline, P11 is located further away and is more influenced by precipitation events

3.4.4. Water balance for the shallow aquifer

Average thickness for the shallow aquifer was estimated at 15 m based on the stratigraphy. The average width of the cross section was estimated at 500 m based on the limits of the mangrove forest (Fig. 3.1). Accumulated precipitation and evaporation were estimated to 145.8 mm and 99.8 mm for the 22 days of the study, respectively.

Table 3.2 summarizes the estimations of groundwater inflow and outflows. Hydraulic gradients (i) were estimated from the piezometric map (Fig. 3.1). Since the saturated zone is located mostly in the alluvium, we use its K estimate to calculate groundwater fluxes. The water balance estimates are shown is Table 3.3. Groundwater flow is the most uncertain estimation since we rely on average aquifer thickness from one cross section and few K estimates.

The positive storage term indicates an increase of about 1.7 mm d^{-1} in the mangrove area. The observed average increase in the water table of shallow wells was 300 mm during a 30 day period.

Table 3.2 Estimated inflow and outflow of groundwater

	i (-)	Ks (m d^{-1})	Cross sectional area (m^2)	Q (m^3 d^{-1})
Groundwater in	0.002	6.7	7500	100.5
Groundwater out	0.003	6.7	7500	167.5

Table 3.3 Summary of the results for the water balance estimation for the period of 21 May to 19 June, 2010

INPUT	Q (m^3 d^{-1})
Precipitation	1325.5
Groundwater	100.5
Surface water	0.0
Subtotal	1426.0
OUTPUT	**Q (m^3 d^{-1})**
Evaporation	907.3
Groundwater	167.5
Surface water	0.0
Subtotal	1074.8
Change in Storage	351.2

3.4.5. Hydrochemistry

Mixing processes between different water types were examined using f$_{sea}$ (Table 3.4), a Piper diagram (Fig. 3.5) and ion correlations (Fig. 3.6). The anion triangle of the Piper diagram shows how samples are grouped into three water types. Shallow wells samples are closely related to the freshwater end-member (Group 1). Shallow wells influenced by grey water (W15O, W17O and W4O) plot between the two end members (Group 2). Nitrate concentrations in the town wells vary between 0.0125 meq L^{-1} and 1.52 meq L^{-1} (Table 3.4), with the highest value observed at well W15O. In contrast, samples taken from the piezometers located in the mangrove forest, show low concentrations of nitrate (0-0.05 meq L^{-1}), because of little grey water influence. Only, P14 shows a distinctive higher nitrate value (0.1326 meq L^{-1}).

Group 3 is clearly more influenced by seawater and corresponds to river outlet samples and samples from piezometers located near to the coast or next to the river. This is supported by the f$_{sea}$ values which vary between 0.04 and 0.4. Highest values correspond to piezometers near the coast, only two of them were sampled (P11 and P14). On the other hand, samples from shallow wells have f$_{sea}$ values between 0 and 0.02.

Table 3.4 Hydrochemistry results in meq L⁻¹, sample locations are shown in Figure 3.1

Sample ID	Source	EC (μS cm⁻¹)	pH (-)	T (C)	Na⁺	K⁺	Mg²⁺	Ca²⁺	Mn²⁺	Fe²⁺	Cl⁻	HCO₃⁻	SO₄²⁻	NO₃⁻	f_sea
RU	SW	569	7.56	27.4	1.01	0.02	0.52	4.74	0.0011	0.0097	0.68	6.60	0.41	0.0077	0.00
P5	GW	2760	7.63	32.8	27.10	0.14	3.16	5.66	0.0660	0.0000	23.88	9.60	1.25	0.0004	0.04
P4	GW	4380	7.68	29.2	35.51	0.14	4.71	11.85	0.0888	0.0016	38.03	9.52	1.98	0.0000	0.07
P11	GW	3230	7.30	28.2	31.90	0.47	3.44	2.83	0.0252	0.0010	26.57	4.20	3.39	0.0509	0.05
Wsw	GW	629	7.57	27.6	1.01	0.01	0.30	5.12	0.0000	0.0000	0.72	6.60	0.19	0.0127	0.00
RO 1	SW	5000	7.61	31.5	46.33	0.69	9.26	5.36	0.0085	0.0000	46.02	6.20	4.75	0.0000	0.09
RO 2	SW	2220	7.74	28.9	15.00	0.29	3.55	5.19	0.0023	0.0014	16.39	6.80	1.89	0.0144	0.03
GrW	grey water	3050	6.97	28.5	31.12	1.33	1.39	5.90	0.0011	0.0000	16.28	18.00	8.31	0.0006	0.03
P14	GW	17230	7.60	28.1	199.71	2.13	34.64	16.42	0.0550	0.0022	204.84	16.40	42.23	0.1326	0.39
P7	GW	7660	7.31	28.1	62.75	0.62	13.07	16.82	0.1822	0.0005	75.46	11.12	4.74	0.0090	0.14
P8	GW	7230	7.47	28.9	39.00	0.66	17.53	27.62	0.5708	0.0003	64.18	6.96	4.96	0.0479	0.12
Sea1	sea	37400	7.60	31.3	460.74	7.40	99.98	15.86	0.0000	0.0003	519.08	2.44	49.99	0.0231	1.00
Sea2	sea	19800	7.36	29.6	217.97	2.09	41.34	8.96	0.0000	0.0000	244.43	4.80	23.16	0.0000	0.47
W15O	GW	2350	7.00	28.1	2.74	0.02	2.05	15.27	0.0001	0.0000	11.77	5.60	2.95	1.5183	0.02
W17O	GW	1386	7.04	28.9	2.12	0.02	1.41	11.21	0.0000	0.0000	6.48	5.20	1.55	0.2846	0.01
W2O	GW	755	7.35	28.1	2.23	0.03	1.03	4.35	0.0000	0.0000	1.86	5.00	0.65	0.0630	0.00
W4O	GW	1528	7.58	28.6	7.35	0.02	2.27	7.60	0.0000	0.0000	8.24	7.40	2.12	0.5037	0.01
W5O	GW	741	7.35	28.5	2.38	0.00	0.79	5.10	0.0000	0.0001	2.40	6.00	0.76	0.0237	0.00
W7O	GW	1015	7.42	28.7	2.72	0.02	0.81	7.26	0.0000	0.0006	2.93	7.80	1.20	0.2692	0.00
W8O	GW	1174	7.47	28.6	4.09	0.03	1.14	8.18	0.0004	0.0004	2.39	8.64	0.94	0.0604	0.00
W9O	GW	876	7.51	28.0	2.39	0.02	0.77	6.66	0.0000	0.0049	1.85	8.68	0.89	0.0125	0.00

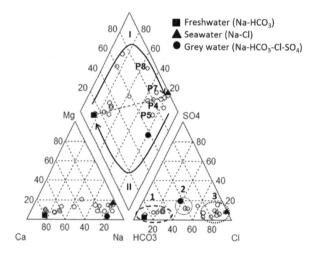

Figure 3.5 Piper diagram including end-members, mixing line (dashed), salinization (I) and refreshening (II) processes. Freshwater end-member (Ca–HCO₃) is sample Wsw. Seawater end-member (Na–Cl) is sample Sea1. Circles in the anion triangle show freshwater samples (1), groundwater with grey water influence (2) and seawater influence (3)

In most cases, samples are grouped closer to the freshwater end-member on the hypothetical mixing line between freshwater and seawater. This is observed in the relationships between major ions and Cl⁻ when compared to the hypothetical mixing line between freshwater and seawater (Appelo and Postma 1993, Ghiglieri et al. 2012). Only P14 and Sea2 are consistently in-between the two end members. Sample Sea2 was taken during low tide and although the Piper diagram classified it as Na–Cl facie, ionic correlation shows mixing with freshwater (Fig. 3.6a). Other piezometer and shallow well samples plot near the freshwater end member. However, P7 and P8 show depletion of Na⁺ relative to Cl⁻ (Fig. 3.6a), and enrichment of Ca²⁺ (Fig. 3.6b), indicating salinization, also shown in the Piper diagram. Whereas P4 and P5 show a slight enrichment in Na⁺ which also confirms the refreshening process indicated in the Piper diagram (Fig. 3.5).

The relationship between Ca^{2+} and Cl^- is characterized by greater dispersion than in the pattern of other ionic correlations. Most of the samples plot above the mixing line, whereas the same samples plot below the mixing line for Na^+ and Cl^-. Additionally, the samples with higher f_{sea} (P7, P8, P4, P14 and Sea2) show a deficit in Mg^{2+} and K^+ (Fig. 3.6c and 3.6d).

Noticeable in Figure 3.6e is the SO_4^{2-} concentration in P14 above the mixing line. Meanwhile, P11 which is also located within the mangrove forest and near the coast, plots very close to the freshwater end member. Other samples plot below the mixing line, indicating SO_4^{2-} depletion.

Manganese was almost absent from shallow well samples and its concentration was very low in river water samples. Nevertheless, Mn^{2+} in piezometer samples within the mangrove forest reached concentrations up to 0.5708 meq L^{-1}. Iron was found at very low concentrations throughout the study area. Nonetheless, concentrations were at least one order of magnitude higher (10^{-3} meq L^{-1}) in piezometer samples than in shallow groundwater samples (10^{-4} meq L^{-1}).

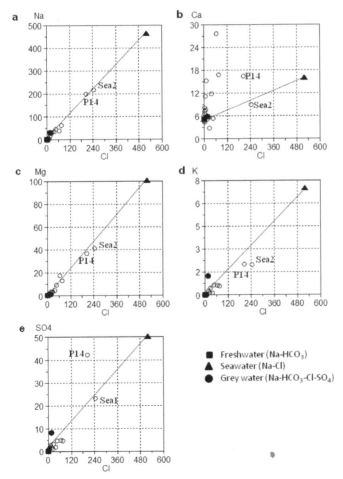

Figure 3.6 Relationship between major ions and Cl^- for all samples. All units in meq L^{-1}

3.5. Discussion

3.5.1. Beach morphology and water balance

The most important source of freshwater during the study period was precipitation, despite high evaporation and groundwater discharge fluxes. Although the evaporation term includes transpiration, direct measurement of the mangrove forest transpiration flux may improve the calculation of the water balance. However, no substantial differences are expected since literature indicates that mangroves conserve water and thus transpiration rates are lower than non-saline plants (Lugo and Snedaker 1974, Saenger 2002).

Sand ridges (Fig. 3.7) are the main geomorphologic features in the study area that control freshwater and seawater mixing and in/outflows. During low flow conditions, the ridges prevent river discharge into the ocean, causing flooding of the mangrove forest and groundwater and surface water accumulation. This is observed as increased groundwater levels. Water accumulation helps to maintain a positive water balance for the mangrove forest during dry periods. Sudden release of water occurs when accumulated water causes rupture of the sand ridges. This only happened once in the study period due to a larger precipitation event (Fig. 3.4). Sand ridges are common on the Pacific Coast of Central America (Jimenez et al. 1999). Several authors describe beach ridges as essential geomorphological features of mangrove ecosystems (Souza Filho and Paradella 2002, Thom 1967, Jeanson et al. 2014); however, their role on the water balance dynamics of mangrove forests has not been studied in detail.

The presence of sand ridges is crucial to guarantee sufficient freshwater availability within the mangrove forest, thus allowing its maintenance during dry periods. This study is among the first to show these dynamic in the Pacific Coast of Central America. Other authors demonstrated that freshwater availability regulates growth, mortality, and phenological events. For instance, changes in hydrological conditions would result in drastic alteration of structural and functional characteristics of the forest (Jimenez 1990, He et al. 2010).

A strong tidal influence was observed in head fluctuations of the coastal piezometers P10 and P12. In contrast, P11 was relatively unaffected by tides. Hydraulic head at Pb11 was around 2 m asl. Since average tidal fluctuation for the Pacific Ocean in Nicaragua is about 2 m (www.ineter.gob.ni), the response in P11 is damped when the head is above 2 m and becomes more evident when the head is below 2 m. Moreover, P11 was located furthest inland and had a f_{sea} of 0.05, which indicates lesser influence of the sea.

3.5.2. Hydrochemistry

Groundwater mineralization is probably caused by calcite dissolution, as has been shown by other authors working in coastal aquifers (Ghiglieri et al. 2012, de Montety et al. 2008, Giménez and Morell 1997). Calcite was found in thin sections from shale samples (Appendix 3.2) indicating possible occurrence of cationic exchange processes which influence the chemistry of groundwater samples.

Coastal piezometers P11 and P14 show significant chemical differences. P11 showed a f_{sea} of 0.05 and its chemical composition resembles that of the freshwater end-member. Moreover, the groundwater level fluctuations show that P11 is not substantially affected by tides. P14 showed a f_{sea} of 0.39, indicating a strong seawater influence on its chemical composition. High concentrations of nitrate in P14 could be caused by inorganic nitrogen from tidal waters, as explained by Rivera-Monroy et al. (1995). In a study of hydrological and hydrogeochemical variations in the Itacurusa Experimental Forest in Brazil, nitrate concentrations peaked with high tide (Kjerfve et al. 1999). This is consistent with the the the Na–Cl facie of this water sample. However, the SO_4^{2-} enrichment with respect to Cl⁻ in this sample cannot be explained merely by freshwater-seawater mixing. Thus, there is a possibility of grey water influence in this sample.

The hydrochemical difference between piezometers on the east side and west side of the river is the result of elevation differences and, thus, groundwater flow directions and fluxes. Piezometers located on the west side of the river showed evidence of refreshening, whereas piezometers on the east side of the river indicated salinization. The deficit in Mg^{2+} and K^+ on the east side is further evidence of salinization process (Walraevens and Van Camp 2005). The clayish soils with high CEC provide the cation exchanger. East side piezometers have a lower hydraulic head than west side piezometers and, consequently, receive more river water contributions.

The hydrochemistry of shallow wells in the town indicate influence of grey water infiltration. The most probable source of sodium and chloride in these wells are laundry and cleaning products (Fetter 1999), which are widely present in grey water. Grey water from households in Ostional is composed of laundry, shower, and kitchen waste waters. Sea water encroachment is ruled out due to the greater distance from the coast, the hydrochemcial evolution from the land towards the sea and the overall groundwater flow direction (towards the sea). Nitrate was also found in some wells (W170, W150 and W40), suggesting that their chemical composition is influenced by the proximity of latrines.

The presence of Mg^{+2} and Fe^{+2} in piezometer samples suggests the occurrence of denitrification within the mangrove forest, following the redox sequence from nitrate to iron reduction. These reduced compounds were not found on samples from the town wells. These results suggest that the mangrove forest is acting as a sink for nitrate in this area. Nitrate is reduced by the high organic matter content of the mangrove sediments. Additionally, highly diverse and abundant microbial communities within mangrove sediments may catalyze the denitrification process (Corredor and Morell 1994). However, further investigation is required to explore this possibility. Mangroves are recognized as nitrate sinks. Denitrification is usually slow in these environments (Alongi 2002, Rivera-Monroy and Twilley 1996), but rates can be increased through nitrate enrichment (Corredor and Morell 1994). Additionally, highly diverse and abundant microbial communities within mangrove sediments may catalyze the denitrification process (Corredor and Morell 1994).

Despite the negative connotation associated with the lack of a sewage treatment system in the study area, this could be also seen as an opportunity for the mangrove forest to receive nutrients inputs. Other authors have pointed out increases in denitrification rates (Corredor and Morell 1994) and in mangrove productivity with sewage inputs (Snedaker 1993). However, there is also evidence that nutrient enrichment by sewage can negatively affect mangrove forests by rising mortality rates, increasing sensitivity to droughts, and affecting tree development due to heavy metal concentration in sewage. Additionally, eutrophication can cause an increase of the greenhouse gas N_2O released by mangrove forests as an intermediate product of microbial denitrification and nitrification (Reef et al. 2010).

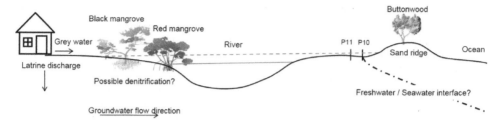

Figure 3.7 Schematic cross-section of the study area, valid for area around cross section B-B'. Dashed line indicates river flooding. For location see Figure 3.1; not to scale. Mangrove tree images from IAN/UMCES image library (http://ian.umces.edu/imagelibrary)

3.6.   Conclusions

A detailed hydrological system analysis applying hydrological, hydrogeological and hydrochemical methods enabled gaining an in–depth understanding of the water balance dynamics in the mangrove forest during a dry season, which is critical to understand the forest subsistence. Precipitation events accounted for the most part of the freshwater input in the system. Net groundwater flux is negative. However, beach ridges along the coast prevent estuarine discharge into the ocean and facilitate storage of surface and groundwater in the mangrove ecosystem. The ridges break when large precipitation events occur, suddenly releasing accumulated water. Freshwater inputs maintain a positive water balance, which may provide adequate conditions for mangrove growth and seedling survival during the dry period. The forest also receives nutrient inputs from the adjacent town. Groundwater mineralization occurs due to calcite dissolution but also due to seawater influence. Refreshening and salinization processes were identified on the west and east side of the river, respectively. These are determined by the general groundwater flow direction. Hydraulic and hydrochemical influence of seawater on groundwater is limited by the water table elevation at the coast and the amplitude of tidal fluctuations. These hydrological and geomorphological characteristics guarantee the continuation of the environmental and socioeconomic services provided by this forest. Future research should focus on the investigation of nutrient inputs from the town and possible denitrification processes within the mangrove forest.

Appendix 3.1 a River estuary during dry season and **b** breach of sand ridges during the rainy season

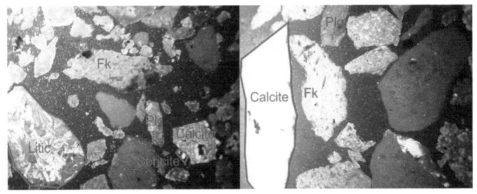

Appendix 3.2 Potassium feldspar (Fk) altered to clay, plagioclase (Plg) altered to carbonates (Calcite) and igneous lithic with plagioclase microlites inside

Appendix 3.3 Shale outcrops at the coast, West of the estuary

Chapter 4

Characterizing the climatic water balance dynamics and different runoff components in a poorly gauged tropical forested catchment

Abstract

The water balance dynamics and runoff components of a tropical forested catchment (46 km^2) in the southwestern Pacific coast of Nicaragua were studied combining hydrometry, geological characterization and hydrochemical and isotopic tracers (3-component hydrograph separation). The climatic water balance was estimated for 2010 /2011, 2011/2012 and 2012/2013 with net values of 811 mm year^{-1}, 782 mm year^{-1} and -447 mm year^{-1}, respectively. Runoff components were studied at different spatial and temporal scales, demonstrating that different sources and temporal contributions are controlled by dominant landscape elements and antecedent rainfall. In forested sub-catchments permeable soils, stratigraphy and steep slopes favor subsurface stormflow generation contributing 50% and 53% to total discharge. At the catchment scale, landscape elements such as smooth slopes, wide valleys, deeper soils and water table allow groundwater recharge during rainfall events. Groundwater dominates the hydrograph (50% of total discharge) under dry prior conditions. However, low soil infiltration capacity generates a larger surface runoff component (42%) under wet prior conditions which dominates total discharge during floods. Our results show that forested areas are important to reduce surface runoff and likely soil degradation which is relevant for the design of water management plans.

Based on: Calderon, H. and Uhlenbrook, S., 2014. Characterising the climatic water balance dynamics and different runoff components in a poorly gauged tropical forested catchment, Nicaragua. Hydrological Sciences Journal. DOI 10.1080/02626667.2014.964244

4.1. Introduction

Sustainable water management requires the broadest possible hydrological information and reliable predictions in order to face water demand for growing populations, prevent ecosystem degradation and reduce the impact of natural hazards and disasters (Sivapalan et al. 2003). Catchment hydrology deeply interweaves with water resources management to ensure life and ecosystem sustainability (Bonell and Bruijnzeel 2004, Uhlenbrook 2006). Nonetheless, data scarcity in ungauged basins poses a major challenge in achieving this understanding.

Prediction of water quantity and quality requires understanding of runoff generation processes (Bonell 1998); and according to Bonell and Bruijnzeel (2004) there has been relatively less research in tropical forests compared to the detailed studies carried out in temperate climates. Runoff generation processes in the tropics are expected to be different from temperate climates due to strong rainfall variability and seasonality. In addition, different soil types and land uses may cause differences between commonly studied temperate regions and poorly investigated tropical areas (Hugenschmidt et al. 2014). Some examples of studies looking at tropical catchments are also found in literature e.g. Elsenbeer *et al.* (1995a), Elsenbeer *et al.* (1995b), Moratti et al. (1997), Elsenbeer and Vertessy (2000), Goller et al. (2005), de Araújo and González Piedra (2009), Roa-García and Weiler (2010), Bruijnzeel (2001), Wenjie et al. (2011), and Munyaneza et al. (2012). Work on runoff generation in tropical climate include Chaves et al. (2008), Germer et al. (2009), Zimmermann et al. (2009), Brooks et al. (2010), Germer et al. (2010), Muñoz-Villers and McDonnell (2013), (Salemi et al. 2013) and Beck et al. (2013).

Key process research studies in Central America include Genereux (2004), who used environmental tracers in Costa Rica to study inter-basin groundwater transfer; Harmon et al. (2009), who studied hydrochemistry of surface water related to geology in Panama; Westerberg et al. (2010), Westerberg et al. (2011) and Guerrero et al. (2012), who studied precipitation and discharge relationships in a catchment in Honduras and another extensive work in Honduras to investigate hydrology and hydrochemistry related to water supply (Caballero 2012). A significant research output for forest hydrology in Costa Rica is listed by Bonell and Bruijnzeel (2004). Some examples are Calvo (1986), Bruijnzeel (2001), Hölscher et al. (2003), Hölscher et al. (2004). Other work in Guatemala and Panama include Holder (2004), Cavelier et al. (1997) and Niedzialek and Ogden (2012). Much of this work is focused on specific aspects of the hydrologic cycle such as interception, throughfall, and stem flow; or on hydrochemistry and nutrient inputs. However, not much work is found in Central America regarding runoff generation processes (Caballero et al. 2013, Häggström et al. 1990). Catchment hydrology studies in Nicaragua are rarely found in the literature e.g. Mendoza et al. (2006).

Noteworthy, Nicaragua is a country with a very modern water resources legislation. The General Law of National Waters (2007) mandates the creation of water resources management plans for each of the 21 catchment areas in the country. However, there has not been yet an official publication of any of these plans. This is probably due to the scarce long term hydrometeorological data monitoring in the country and other resources related challenges.

The South Pacific coastal area of Nicaragua is a zone of great touristic potential and real estate development is taking off quickly. The investigated Ostional catchment is located within this area and has an extraordinary coastal and rural touristic potential. Nevertheless, increase in tourism and other related developments will create large stress on water resources in this region. For instance, the main town water supply is based on a single deep well that runs already now at its maximum capacity. Therefore, sustainable development of water resources in the area requires knowledge of water resources availability and understanding of the hydrological system.

The study area provides a typical example of a poorly gauged, tropical forested catchment on the South Pacific Coast of Nicaragua. This catchment provides an opportunity for testing methods to investigate rainfall–runoff processes commonly used in temperate climates in a different hydro-climatic region as suggested by Burns (2002). Additionally, this study covers different spatial (sub-catchment and catchment) and temporal scales (rainy season and individual rainfall events). This approach enables analysis of the effect of different catchment structural characteristics and hydro-climatic controls on runoff generation processes.

The objective of this research was to calculate the climatic water balance for the catchment and to understand the main runoff generation sources and flow paths by applying different hydrochemical and isotopic tracers at different spatial and temporal scales.

This is one of the first works in a forested catchment in the Central American region which uses a combined approach of hydrometry, hydrological characterization (based on geology, geomorphology, topography and land use) as well as chemical and isotopic tracers. Therefore, the gained knowledge not only provides information on water resources availability for developing water management plans, but also provides a reference for future studies in similar hydrological settings in the region and intercomparison with other catchments in the world.

4.2. Study area

The Ostional catchment is located at the Southwestern coast of Nicaragua. It has an approximate area of 46 km^2. Two sub-catchments were selected for event based sampling: Rompeviento (0.8 km^2) and El Nancite (1.6 km^2). The selection was based on proximity to a rain gauge station and accessibility. Locations are shown in Figure 4.1.

According to Holdrige's classification (Holdridge 1967), the area is located in the tropical wet forest life zone. The rainy season starts in May and ends in November. September and October are usually the rainiest months at the Pacific Coast. Usually between July and August, there is a 3 to 5 week period known as midsummer drought (MSD) (Magaña et al. 1999). Historical climatic data for the period 1965–2007 was available for Rivas station, located 40 km NW of the catchment and at an elevation of 70 m asl. (CIRA 2008). The 42 year record registered a mean temperature of 27.1°C, a monthly minimum of 24.2 °C registered in February and a monthly maximum of 30.6°C registered in May. Wind direction is predominantly towards the East and average wind velocity is 5 m s^{-1}. Historical mean annual precipitation at Rivas is 1,476 mm year^{-1} and mean pan evaporation was 1,976 mm year^{-1}.

Elevation ranges between sea level and 500 m asl. in the NE. Land use in 2002 was dominated by forest (52%), agriculture (20%) and pasture (28%) (UNA 2003). Soil depths vary between 0.9–0.5 m and texture is classified as clay-loam and clay (UNA 2003). The study area is located in the geomorphologic province of the Pacific Coast, in the Rivas–Tamarindo sub-province. The region developed from sediment deposition during sea regression and transgression. Two geologic Formations are present in the area: the Upper Cretaceous Rivas Formation and the Eocene Brito Formation. However, only the Brito

Formation is exposed. The Rivas formation is composed of drab, tuffaceous shale, sandstone, arkose and greywacke (Swain 1966). The Brito formation, overlies concordantly the older Rivas formation. The Brito formation is comprised of volcanic breccias, tuffs, shales, limestones and sandstone (Krasny and Hecht 1998, Swain 1966, Elming et al. 2001).

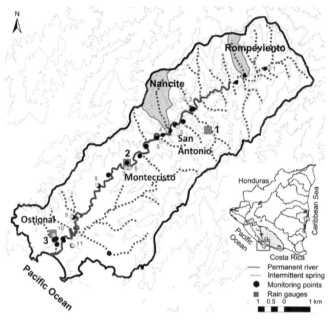

Figure 4.1 The Ostional catchment and selected sub-catchments (Nancite and Rompeviento). Black circles are groundwater level monitoring points for the period 2010-2013 and rain gauge locations: 1) Monteverde, 2) Montecristo and 3) Ostional. Gray circles are locations of soil infiltration tests. Topographic contour resolution is 100 m

4.3. Materials and methods

4.3.1. Hydrological monitoring

The monitoring period spanned from May 2010 to April 2013. Three automated rain gauges (Rain–O–Matic Professional rain gauge from PRONAMIC) were installed to collect rainfall data with a resolution of 0.254 mm. Monitoring sites were selected based on elevation differences and accessibility. Locations are presented in Figure 4.1.

Class A evaporation pans were installed at the same locations to calculate reference evaporation. The pan evaporation is related to the reference evaporation by an empirically derived pan coefficient depending on site specific conditions such as pan type, land cover in the station, its surroundings, as well as the general wind and humidity conditions. The term reference evaporation is defined as the rate of combined evaporation and transpiration from a reference surface, with no water shortage. This term allows estimation of total evaporation demands of the atmosphere independently of crop type, crop development and management practices (Allen et al. 1998). The pan evaporation is related to the reference evaporation by an empirically derived pan coefficient according to Equation 4.1 (Allen et al. 1998):

$$E = K_p \, E_p \tag{4.1}$$

where E is the reference evaporation, E_p is pan evaporation, both in mm day^{-1}; and K_p is the pan coefficient (-), which depends on site specific conditions such as pan type, land cover under the evaporation pan, type of crop surrounding the pan, general wind speed and air humidity conditions. In our case, the coefficient was estimated at 0.7. Measurements of precipitation and evaporation were taken every morning.

Missing rainfall and evaporation data were filled in by the normal ratio method. This method estimates missing rainfall at the station under consideration as the weighted average of adjoining stations. The rainfall at each of the adjoining stations is weighted by the ratio of the average annual rainfall at the station under consideration and average annual rainfall of the adjoining station (Paulhus and Kohler 1952) using the other available stations data. Total rainfall and evaporation for the catchment was estimated using the weighted average from the three monitoring stations using Thiessen polygons.

Groundwater levels were measured monthly using a water level meter (Solinst, model 102) in all shallow household wells in the three communities of the catchment for the period of July 2010 to April 2013. Well depth in all locations ranges between 4 and 14 m. The wells in the communities of San Antonio and Montecristo, in the upper catchment area, are located near the river. The wells in the Ostional community are more scattered (Fig. 4.1).

Soil infiltration capacity was measured using double ring infiltrometer tests (Rawls et al. 1996). Eleven sites were selected along the catchment based on different soil types and slopes; locations are shown in Figure 4.1.

River discharge was estimated by developing a rating curve through the area–velocity method (Buchanan and Somers 1969). Velocity was measured using a flow meter from Global Water FP111. Frequent measurements (218 in total) were performed at the catchment outlet in the Ostional town starting in May 2010. River stage was monitored using a Schlumberger diver, range 10 m, DI501. Data was recorded every 30 min. Barometric pressure was recorded at the same interval and period to compensate for atmospheric pressure variations using a baro diver, range 1.5 m (Schlumberger 50013, DI501) located in the catchment area.

Attempts to build and update the rating curve for the catchment were made during the study period, but it was not possible to carry out continuous measurements due to technical problems. However, discharge–stage (Q–h) analyses could be performed for nine rainfall–runoff events during 2012. During these events, most of the rating curves yielded a correlation coefficient (R^2) of less than 0.2, likely caused by the changing channel morphological conditions during the events. Only two events yielded R^2 values above 0.8: an event on the 19 October 2012 (R^2=0.82) and an event on the 22 October 2012 (R^2=0.84).

4.3.2. Hydrochemical and isotopic sampling

Bulk rain water samples were collected weekly during the rainy season of 2010 from each rain gauge station for isotopic analysis and construction of the local meteoric water line. A total of 80 samples were collected.

Event based sampling was carried out at sub-catchment scale in 2010 at the outlets of the Nancite and Romepeviento; and at catchment scale in 2012 at the outlet of Ostional, near the coast. Sampling frequency was 5 min and 15 min for sub-catchment scale events and 20 min for catchment scale events. Electrical conductivity (EC), pH, stream velocity and stage were recorded at each sampling time. Samples were immediately filtered through 0.45 μm glass fiber filters. The samples used for cations (25 mL) analyses were acidified with concentrated H_2NO_3 to prevent precipitation reactions and cation attachment to the surface of the sample bottle. The samples for anions (50 mL) were stored in non-acidified bottles. Samples for isotopic analyses of δ^2H and $\delta^{18}O$ were collected in 1 ml glass vials.

In this study, rainfall events were defined as periods of continuous precipitation yielding at least 23 mm d^{-1}. A total of 4 events were sampled for sub-catchment scale and 9 for catchment scale. However, only 2 and 3, for sub-catchment and catchment scale, respectively; showed a recognizable peak in discharge. The other events were too small to show significant chemical or isotopic changes.

Major cations (Mg^{2+}, Na^{2+}, Mn^{2+}, K^+, and Fe^{2+}) were analyzed using an Inductively Coupled Plasma-Optical Emission Spectrometer (ICP-OES: Perkin Elmer Optima 3000); while Ca^{2+} was measured using a flame Atomic Emission Spectroscope (AES: Perkin Elmer AAnalyst 200). The major anions (Cl^-, HCO_3^-, SO_4^{2-}, NO_3^{2-}) were measured using an Ion Chromatography System (IC: Dionex ICS 1000). The calculated ion balance error was between -10% and +10%; samples with higher errors were discarded. SiO_2 was analyzed using the molybdosilicate method in a spectrophotometer (Clesceri et al. 1998). The detection limit for the method is 0 to 1.71 mg SiO_2/50 ml). δ^2H and $\delta^{18}O$ were measured in an LGR Liquid Water Isotope Analyzer (LWIA). Analytical error for δ^2H amounts to ±1‰ and for $\delta^{18}O$ to ±0.2‰. All the analyses were performed at the hydrochemical laboratory at UNESCO-IHE in the Netherlands.

4.3.3. Hydrograph separation

Total runoff was analyzed using three-component hydrograph separation as outlined by Hoeg et al. (2000). This method is based on the mass conservation of water and the tracers (Sklash and Farvolden 1979). Chemical tracers were used to determine source areas contributing to total runoff and water isotopes were used to determine temporal sources of runoff as described by other authors (Uhlenbrook et al. 2002, Wenninger et al. 2004, Munyaneza et al. 2012).

At sub-catchment scale events we were able to use Mg^+, Na^+ and Cl^-. However, at catchment scale we used Cl^- and silica for the three-component separation since they behaved more conservatively based on the chemographs which show highly irregular patterns for other tracers. Capturing the spatial and temporal variability of end-members is difficult as discussed by other authors (Uhlenbrook and Hoeg 2003, Didszun and Uhlenbrook 2008, Uhlenbrook and Leibundgut 2002, Blume et al. 2008a, Bohté et al. 2010, Hugenschmidt et al. 2010). Silica has proven to be an appropriate tracer for hydrograph separations (Hugenschmidt et al. 2014, Hoeg et al. 2000, Blume et al. 2008b, Mul et al. 2008, Blume et al. 2008a). Munyaneza et al. (2012) used chloride and silica to perform two-component separations.

Selected end-members for the sub-catchment scale events were surface runoff (event water), based on a bulk rainwater sample collected during one of the events; groundwater, based on average chemical composition of shallow wells during the dry season; and subsurface stormflow, based on a sample from a shallow well (5 m deep) located at the base of the hillslope in the Nancite sub-catchment. The sample was collected after the rainy season, and we assume that this shallow well collects subsurface stormflow from the hillside. A similar assumption was used by Hooper et al. (1990).

Selected end-members at the catchment scale were: surface runoff, baseflow and groundwater. The surface runoff end-member is based on the average composition of rainwater. Baseflow is based on river water collected after a 10 days period with an accumulated precipitation of 25 mm in the catchment. Here we use the definition of Dingman

(2002) of baseflow, as flow that cannot be associated with a specific event. The small amount of precipitation indicates negligible contributions from event-associated runoff. Groundwater is based on the average composition of shallow groundwater along the catchment during the dry season. Runoff coefficients were determined for each event based on the ratio of event discharge over total precipitation as discussed by Blume et al. (2007).

4.4. Results

4.4.1. Hydrological characterization

Fifteen intermittent and perennial first order springs were found in the medium and upper part of the catchment (Fig. 4.1). Flow rates during dry season vary between 0.005 m^3 s^{-1} and 1 m^3 s^{-1}; streams located at the uppermost catchment area have the higher discharge rates.

Catchment geology in the upper catchment area is defined by massive sandstone, overlaid by highly folded and fractured shale and limestone. Stream discharge is sustained by groundwater as confirmed by the abundance of travertine deposits near the stream headwaters (Appendix 4.1). Springs occur at the contact between the shale/limestone and the sandstone. We observed in the field that sandstone is less deformed and much less permeable than the shale unit. Therefore, springs occur along the faults in the shale/limestone or at the contact with the less permeable sandstone. Densely distributed joints perpendicular to the fractures in the shale and limestone favor preferential flow.

The sub-catchments are narrow V–shaped valleys (Appendix 4.2a), where alluvial deposition only reaches about 1 m in thickness and stream width is between 1–2 m. Sandstone streambeds are exposed and highly fractured shale walls (10–20 m) enclose the streams. Population in this area is very scattered and land use is dominated by natural forest. Slopes are very steep, between 20% and 100%.

The lower part of the catchment presents a wide (1 km) alluvial valley (Appendix 4.2b) with an estimated thickness of up to 15 m. Underlying the alluvium, a shale unit of unknown thickness was found (Calderon et al. 2014). Slope is less than 1%. Land use in the lower catchment is a combination of forest, agriculture, and pasture. Soil infiltration capacity values vary between 10 mm h^{-1} to 50 mm h^{-1} (Table 4.1). Higher values correspond to the upper catchment area and lower values correspond to the lower part of the catchment.

Table 4.1 Soil infiltration capacity

Site	Infiltration capacity (mm h^{-1})
1	50
2	48
3	40
4	45
5	36
6	13
7	18
8	36
9	15
10	10
11	10

4.4.1.1. Rainfall and evaporation

Monthly rainfall values are presented for the period of May 2010–May 2013 (Fig. 4.2). For all stations rainfall during 2010 was more evenly distributed during the rainy period compared to 2010 and 2011, when large peaks were observed during October 2010 (Fig. 4.2a). Annual rainfall for Ostional, Montecristo and Monteverde was 2,049 mm year^{-1}, 2,243 mm year^{-1} and 2,083 mm year^{-1} for 2010; 1,636 mm year^{-1}, 2,087 mm year^{-1} and 2,358 mm year^{-1} for 2011; and 1,581 mm year^{-1}, 997 mm year^{-1} and 1,094 mm year^{-1} for 2012, respectively. Rivas station registered 1,684 mm year^{-1}, 1,893 mm year^{-1} and 1,018 mm year^{-1} for 2010, 2011 and 2012, respectively.

In all cases, 2012 was a relatively dry year compared to the historical mean annual value (1,476 mm year^{-1}). Extremely high values were observed for October 2011, between 1,000 mm month^{-1} and 1,300 mm month^{-1}. The reference station Rivas also showed a high value, *i.e.* 600 mm month^{-1}, the double of the historical mean. A significant spatial variation in rainfall within the catchment was observed. During 2010 difference in accumulated rainfall between stations varied between 2% and 9%. For 2011 differences were between 12% and 44% and for 2012 between 9% and 36%. In 2012, rainfall was highest near the coast, which indicates influence of sea evaporation. Peak evaporation values were observed during March 2011 with 170 mm month^{-1} for the catchment (Fig. 4.2b).

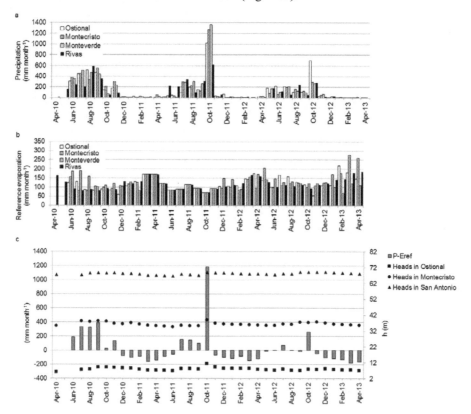

Figure 4.2 Climatic water balance for the catchment for the period June 2010 to April 2013. **a** Monthly precipitation (P) for each station. **b** Reference evaporation (E$_{ref}$) at each station. **c** P–E$_{ref}$ for the catchment (bars) based on area weighted averages and average hydraulic head (h) fluctuations for wells at the three catchment communities

4.4.1.2.   Climatic water balance

For all wells during the three years a slow increase in head is observed with the onset of the rainy season. Heads stay high for about two months after the end of the rainy season, suggesting groundwater recharge and storage, and then start to gradually fall again. The rainy season of 2011 was an exception. Since most precipitation was concentrated in October, sudden short rises in the hydraulic heads at the three communities were observed during this month. During 2010 and 2012, we observed a more even distribution of precipitation during the rainy period, which caused a smoother rise and fall in hydraulic heads within a period of two months.

Average groundwater level fluctuations were larger for the lower catchment wells (Ostional), with increases of 33%, 41% and 12% between the dry and wet season for each monitored year. Average increases for the middle catchment wells (Montecristo) were 8%, 11% and 2%. For the upper catchment wells (San Antonio) average increases were the lowest: 2%, 3% and 2% for each year. Fluctuations between wells are larger than within wells. For instance, maximum differences within wells were 9%, 22% and 57% for San Antonio, Montecristo and Ostional, respectively, during the total monitoring time. Differences between wells were 30%, 60% and 80%, for San Antonio, Montecristo and Ostional, respectively.

A net positive balance was estimated for the hydrologic years 2010/2011 and 2011/2012 with values of 811 mm and 782 mm, respectively. For 2012/2013 the balance was negative (-447 mm).

4.4.2. Rainfall-runoff events

4.4.2.1.   Hysteretic behaviour of solute concentration-discharge relationships

Hysteresis in tracer concentration and discharge relationships were analyzed. All tracers yielded similar patterns; therefore, only EC is shown for discussion (Fig. 4.3). Clockwise hysteresis loops were observed for all events at sub-catchment and catchment events, except for the 22 October event. Clockwise patterns indicate higher concentrations during the rising limb of the events (House and Warwick 1998, Burt et al. 1983). Counter-clockwise patterns show lower solute concentrations during the rising limb of the hydrograph. Higher concentrations of tracers in the rising limb may indicate contribution from delayed subsurface runoff which may show little tracer dilution during storms (Burt et al. 1983, House and Warwick 1998). The 22 October event showed increased concentrations in the falling limb of the hydrograph compared to the initial concentrations. For instance, Ca^{2+} increased from 43 mg L^{-1} to 99 mg L^{-1}; Na$^+$ from 6.8 mg L^{-1} to 15.7 mg L^{-1}; and Mg^{2+} from 1.98 mg L^{-1} to 4.4 mg L^{-1}.

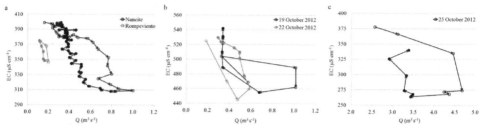

Figure 4.3 Hysteresis in EC–discharge relationships during **a** sub-catchment and **b, c** catchment rainfall–runoff events. Hollow circles represent concentrations during the rising limbs, solid circles represent concentrations during the falling limbs, arrows indicate time evolution

4.4.2.2.   Hydrograph separations

Open rainfall samples were used to determine surface runoff composition for catchment scale events. Therefore, to account for enrichment caused by throughfall, a theoretical enrichment of 1.4‰ was assumed for both water isotopes. Isotopic enrichment by throughfall has been reported in forested catchments in Canada (+2.9‰), Japan (+2.8‰), Germany (+36‰) and Costa Rica (+1.4‰) (Klaus and McDonnell 2013, Rhodes et al. 2006). We also compared the composition of an open rainfall sample and under–canopy rainfall sample for a sub-catchment event. We found an enrichment of +1.5‰ in the under–canopy sample with respect to the open rainfall sample.

Table 4.2 End-member composition for sub-catchment and catchment scale events

Scale	End-member	SiO_2 (mg L^{-1})	Cl^- (mg L^{-1})	$\delta^{18}O$ (‰)	δ^2H (‰)
	Surface runoff	na	5.74	-7.40	-52.10
Sub-catchment	Subsurface stormflow	na	16.11	-7.13	-40.80
	Groundwater	na	12.09	-6.34	-40.98
	Surface runoff	0.051	2.48	-7.45	-56.57
Catchment	Groundwater	0.748	18.91	-6.70	-43.30
	Baseflow	1.060	5.96	-6.27	-36.10

Sub-catchment scale (Evenst 1 and 2) Total contributions from subsurface stormflow, groundwater, and surface runoff were 50%, 35% and 15% for Event 1, respectively; and 53%, 28% and 19% for Event 2, respectively.

During Event 1 (Rompeviento) subsurface stormflow was almost equal to groundwater discharge and then increased with the peak of the hydrograph (Fig. 4.4). In the case of Event 2 (El Nancite), subsurface stormflow accounted for the total discharge at the beginning of the hydrograph. Groundwater contribution was approximately constant throughout Event 1. Peak groundwater contribution occurred at 11:45 LT, right before peak discharge. At this moment, groundwater discharge decreased and surface runoff increased. A similar behavior is observed in Event 2. Groundwater contribution was the highest during peak discharge at 17:22 LT and surface runoff peaked when groundwater contribution decreased.

62

Isotopic hydrograph separation results were different for $\delta^{18}O$ and δ^2H, although usually information from these stable isotopes is considered similar. The hydrograph separation based on $\delta^{18}O$ determined that in Event 1 pre-event water contribution was 42% and in Event 2 the contribution was 46%. However, the hydrograph separation using δ^2H shows a larger contribution of pre-event water. For event 1 is 75% and for event 2 is 58%.

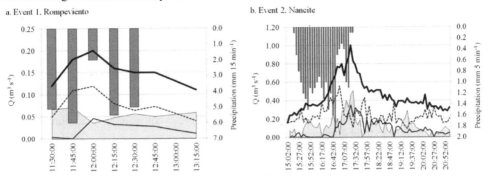

Figure 4.4 Hydrograph separation based on chloride and $\delta^{18}O$ for Events 1 and 2 (sub-catchment scale). Q_t: total discharge, Q_{sr}: surface runoff, Q_{ssf}: subsurface stormflow and Q_{gw}: groundwater

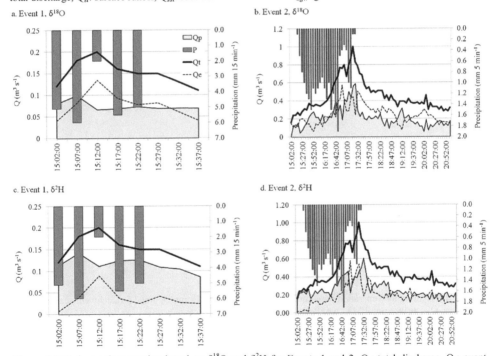

Figure 4.5 Hydrograph separation based on $\delta^{18}O$ and δ^2H for Events 1 and 2. Q_t: total discharge, Q_e: event water contribution and Q_p: pre-event water contribution

  Catchment scale (Events 3, 4 and 5) Several ions were tested to perform hydrograph separations, but Cl⁻ and $\delta^{18}O$ were selected since they behaved more conservatively. Other tracers were also tested to separate the hydrographs, but the mass balance was not closed. Figure 4.6 presents results for three events. Chemical hydrograph separation for Event 3 (19 October 2012) showed contribution of surface runoff, baseflow and groundwater of 19%, 36% and 45%, respectively. For Event 4 (22 October 2012) contributions in the same order were 32%, 22% and 46% and for Event 5 (25 October 2012) 42%, 38% and 21%, respectively.

  Similar results were obtained using either water isotope (Fig. 4.7). Pre-event water contributions based on $\delta^{18}O$ were 82%, 68% and 70% for Events 3, 4 and 5, respectively. Contributions based on $\delta^{2}H$ were 84%, 82% and 66% for the same events. Only for Event 4 the difference between results from $\delta^{18}O$ and $\delta^{2}H$ is large. This result may be caused by a different event type (Lyon et al. 2009). Nevertheless, pre-event water was accounted as the major contributor to total discharge for both isotopes.

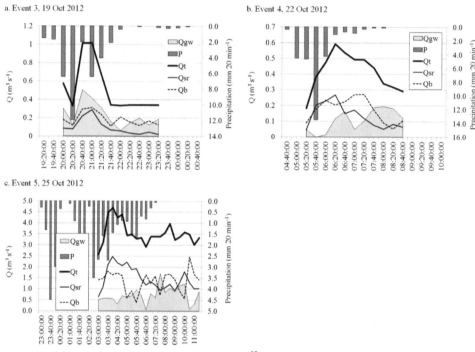

Figure 4.6 Hydrograph separation based on chloride and $\delta^{18}O$ for Events 3, 4 and 5 (catchment scale). Q_t: total discharge, Q_{sr}: surface runoff, Q_b: baseflow and Q_{gw}: groundwater

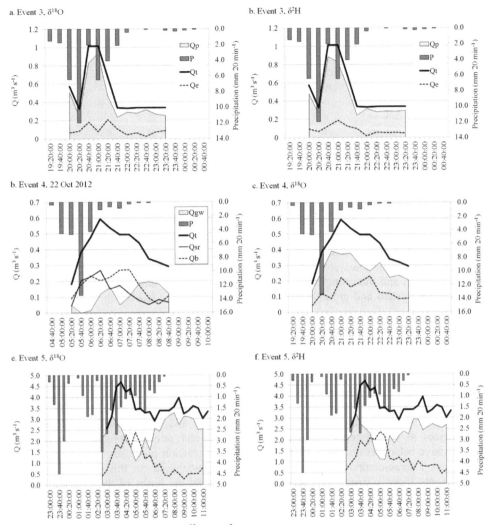

Figure 4.7 Hydrograph separation based $\delta^{18}O$ and δ^2H for Events 3, 4 and 5. Q_t: total discharge, Q_e: event water contribution and Q_p: pre-event water contribution

4.4.3. Runoff coefficients

Runoff coefficients were low (Table 4.3), while the largest value was estimated for Event 2. In the case of sub-catchment events, rainfall intensity was below the highest infiltration capacity of the soils (50 mm h^{-1}). For catchment scale events, rainfall intensity was above the infiltration capacity of soils at the lower catchment area (10 mm h^{-1}).

Table 4.3 Summary of rainfall-runoff events. Runoff coefficients are based on $\delta^{18}O$ hydrograph separations

Event	Scale	Area	Date	Rainfall duration	Max rainfall intensity	Rainfall amount	Peak discharge	Event runoff	Runoff coefficient
		(km^2)		(h)	$(mm\ h^{-1})$	(mm)	$(m^3\ s^{-1})$	(mm)	(%)
1 (Rompeviento)	sub-catchment	0.8	15 August 2010	1.5	24.0	23.60	0.20	0.32	2.80
2 (Nancite)	sub-catchment	1.6	16 August 2010	2.4	22.8	25.20	1.00	2.57	12.98
3	catchment	46	19 October 2012	2.0	35.9	37.00	1.01	0.02	0.07
4	catchment	46	22 October 2012	4.0	40.5	30.72	0.59	0.02	0.12
5	catchment	46	25 October 2012	8.3	13.5	34.36	4.68	0.68	1.72

4.5. Discussion

The climatic water balance was positive for the hydrologic years 2010/2011 and 2011/2012, but it was negative for 2012/2013. However, we observed baseflow during the dry season 2012/13 (Appendix 4.3). Therefore, we infer that groundwater storage sustained river discharge during dry months. Unfortunately, it was not possible to estimate the catchment water balance continuously, due to technical problems related to the construction of the gauging station and the rating curve. Therefore, we were not able to separate groundwater stored in the system and groundwater discharged as baseflow leaving the catchment to the Pacific Ocean.

Selection of three-component over two-component hydrograph separation was based on the analysis of hysteretic patters for chemical tracers. Analysis of hysteresis loops for the Rompeviento sub-catchment event show a small shift between the rising and falling limb. The fact that the shift is small suggests that a two-component model is appropriate to explain total runoff composition for this event. Since a simple mix of two end-members would result in an almost collinear EC-discharge relationship (cf. Sklash and Farvolden, 1979). Larger shifts were observed in the relationship between EC and Na^+ as well as EC and Cl^-. Differences in the hysteresis patterns between EC and selected ions indicate different conservative behaviors and thus different transport or reaction processes (Hoeg et al. 2000). Therefore, a three-component hydrograph separation was used for this event. Catchment scale events show a larger shift between rising and falling limbs of the hysteresis loops. Therefore, three-component hydrograph separation was also more appropriate for these events (Sklash and Farvolden 1979, House and Warwick 1998, Hoeg et al. 2000).

4.5.1. Sub-catchment scale

Hydrograph separations at sub-catchment scale indicate that rainfall infiltrates quickly into the thin soils of the hillslopes activating subsurface stormflow. We did not observe surface runoff during these events. Macropores from the roots of the abundant vegetation, and probably insect and animal burrows, favor infiltration over surface runoff despite the steep slopes. Subsurface stormflow occurs at the contact between the permeable soil and the less permeable shale/limestone rocks, as we observed in the field. Thus, subsurface stormflow is probably delivered to the stream through the soil matrix and macropores/pipes processes as discussed by others (Weiler et al. 2006, Bonell 1993, McDonnell 1990). Detailed process examination was beyond the scope of this study, as this would need more intense small scale investigations including soil moisture patterns.

Groundwater discharge is activated when rainfall percolates beyond the soil layer probably causing a rise in the water table, which is already near the streambead (<1 m bgl). At this point, subsurface stormflow decreases and groundwater becomes a larger contributor to total runoff. Although the riparian zone in these sub-catchments is only about 4 m wide, we believe that the shallow water table and the fractures in the shale and limestone (Krasny and Hecht 1998), which may induce preferential groundwater flow, enhance groundwater contribution to total discharge during rainfall–runoff events.

Hydrographs dominated by subsurface runoff components have been also observed in the work of others, which is mostly concentrated in temperate climates (Sklash and Farvolden 1979, Capell et al. 2011, Cey et al. 1998, Didszun and Uhlenbrook 2008, Wels et al. 1991, Genereux and Hooper 1998), but also in tropical semi–arid (Bohté et al. 2010, Hrachowitz et al. 2011) and humid tropical (Hugenschmidt et al. 2010) climates. Although other work in tropical catchments in the Amazonas (Elsenbeer and Vertessy 2000, Elsenbeer et al. 1995a, Elsenbeer 2001), Australia (Elsenbeer et al. 1995b) and Puerto Rico (Schellekens et al. 2004) reported overland flow as the dominant component of storm runoff.

Differences in the hydrograph separations based on $\delta^{18}O$ and δ^2H are explained by various sources of errors in the end member concentrations (cf. Uhlenbrook and Hoeg, 2003) and analytical errors. However, Lyon et al. (2009) indicate that there are independent variations in sampled precipitation values for both isotopes, which reflect various sources of atmospheric moisture generating rainfall. In our case, the difference in the stable isotope separations may have been caused by the spatial isotopic variability of rainfall, which is not captured by the selected end-member.

4.5.2. Catchment scale

Antecedent rainfall and soil saturation appear to be the most important factors in determining the contribution of surface runoff to total discharge during rainfall events. This is supported by the fact that Event 3 shows the smaller contribution to total discharge than Events 4 and 5. During Event 3, groundwater was the most important component. For Events 4 and 5 the beginning of the hydrograph is dominated by surface runoff and baseflow became the dominant component after peak discharge.

The differences in the dominant component are related to antecedent rainfall. Prior to Event 3, there were no large enough rainfall events to produce a significant increase in total discharge. This indicates that the soils were dry enough to infiltrate the rainfall which occurred during Event 3; whereas during Events 4 and 5, soils must have been close to saturation.

Groundwater contribution during Event 3 was the largest during peak discharge and then decreased once the rainfall input ceased. This suggests that infiltration caused a rise in the water table which must have been already near the river bed to respond quickly to the rainfall event. For Events 4 and 5, groundwater contribution was delayed with respect to peak discharge. This suggests that the water table had to be replenished before there could be significant groundwater contribution.

Hydrograph separation based on stable water isotopes indicate that in all events pre-event water was the main component of the hydrograph. Interestingly, during Event 5 there was a shift from pre-event water to event water at 5:00 LT. This result agrees with the chemical separation which indicated dominance of the surface runoff component.

4.5.3. Runoff coefficients

Small runoff coefficients for sub-catchment events agree with the chemical hydrograph separation which indicates a larger subsurface flow contributions (stormflow and groundwater). However, there is a discrepancy between $\delta^{18}O$ and $\delta^{2}H$ results. This discrepancy limits the applicability of the runoff coefficient.

We observe the smallest runoff coefficient at catchment level for Event 3 and somewhat larger for Events 4 and 5. The difference between events agrees with the small surface runoff contribution estimated for Event 3 and the larger contribution of this component to Events 4 and 5. Runoff coefficients estimated a smaller amount of event water compared to the estimated surface runoff (-37%, -34% and -53% for Events 3, 4 and 5).

4.5.4. Synthesis: Conceptual model of runoff generation

Sub-catchments are characterized by steep slopes, thin clayish soils overlying shale/limestone, shallow water table and narrow riparian zone (Fig. 4.8a). Intermittent springs are formed by (a convergence of) faults and preferential flow of groundwater through perpendicular joints and fissures. The dominant runoff component was subsurface stormflow, followed by groundwater and surface runoff. Macropores (fissures, root channels etc.) in clay–shale/limestone slopes may produce rapid water percolation and subsoil drainage (Bogaard et al. 2012). Small runoff coefficients (2.8% and 12.98%) also indicate little generation of surface runoff. Separation between event and pre-event water was not conclusive.

At the catchment scale (Fig. 4.8b), we find an approximately 1 km wide alluvial valley with an estimated thickness of 15 m (Calderon et al. 2014). Land use is more diverse, including pasture and agriculture as opposed to the forested sub-catchments. At this scale, the composition of total runoff is influenced by antecedent rainfall events. Groundwater dominates in dry soil conditions and surface runoff is the dominant component when soils are (nearly) saturated. This agrees with the dominant pre-event water component during drier conditions and dominant event-water component during wetter conditions. Runoff coefficients also increase (from 0.07% to 1.72%) with antecedent rainfall.

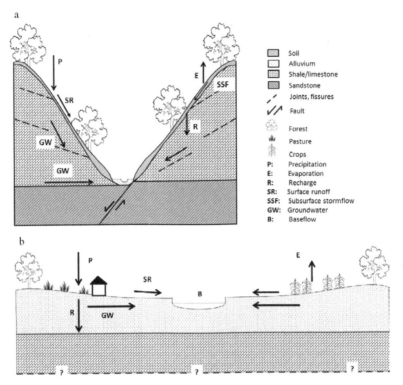

Figure 4.8 Conceptualization of runoff generation at **a** sub-catchment and **b** catchment scale

4.6. Conclusions

The net climatic water balance for the study period was positive. Groundwater recharge from rainfall increases water table levels and maintains baseflow during dry periods. Unfortunately, changing morphological conditions at the river channel impeded continuous estimation of surface water discharge and, consequently, complete water balance dynamics.

Runoff composition changes according to landscape characteristics such as catchment size, slope, geology, stratigraphy and land use. In forested sub-catchments permeable soils, stratigraphy and steep slopes favor subsurface stormflow generation. At catchment scale, smooth slope, deeper soils and water table allow groundwater recharge during rainfall events. Total runoff is dominated by groundwater and baseflow under dry prior conditions. However, low soil infiltration capacity generates a larger surface runoff component under wet antecedent conditions.

The combined use of hydrometric tools with chemical and isotopic tracers in an integrated fashion is a novel approach for runoff generation research in tropical regions. Our results show that forested areas are important to reduce surface runoff and, thus, soil degradation. This is relevant for the design of water management plans in the study area.

Appendix 4.1 Travertine deposits encountered during the dry season 2011 at the headwaters of the catchment

Appendix 4.2 Different landscape characteristics at sub-catchment (a) and catchment (b) level

Appendix 4.3 River baseflow during dry season 2012 in the upper catchment area

Chapter 5

Lessons learned from catchment scale tracer tests during rainfall–runoff events in a tropical environment using natural DNA from total bacteria and qPCR

Abstract

Environmental applications of natural and artificial DNA using quantitative Polymerase Chain Reaction (qPCR) are extensively reported in literature. However, despite the advantages of the low detection limit of qPCR and the possibility of an unlimited number of DNA tracers that can be used, very few hydrological applications are reported and no use of natural DNA as a tracer has been reported yet. Use of artificial DNA tracers have been used in small scale (100 m - 1000 m) field sites to investigate transport parameters. Our work reports a field scale (11000 m) test of natural occurring bacterial DNA as a tracer during rainfall-runoff events. The objective was to test the applicability of natural DNA from total bacteria to separate the total runoff signal into its hydrological components. Synoptic sampling throughout the catchment was performed to determine background bacterial DNA content. The study area is a tropical forested catchments with clayish soils located on the southwestern Pacific coast of Nicaragua. Inhibitory substances present in surface runoff contributions to stream affect DNA amplification during qPCR. This is observed in the inhibition of qPCR for surface water samples during the rainy season. Groundwater samples collected in this period showed no qPCR inhibition, but bacterial content decreased probably due to dilution from local precipitation. Sample dilution combined with the use of bovine serum albumina (BSA) in the qPCR mix solves the inhibition issue. However, the optimal concentration of BSA should be further investigated. The DNA harvesting method used in situ was successful. Nonetheless, DNA losses during the pre-filtration step have to be evaluated. This is a promising technique for hydrological research, but more field scale experiments are required to use bacterial DNA to investigate rainfall-runoff processes in a quantitative way. DNA recovery and qPCR inhibition in runoff samples have to be addressed in future works. Future experimentation should include areas with different soil and land use types. Although it was not possible to quantify the bacterial content in the samples, we can draw valuable lessons for future works.

This is an extended version of: Calderon, H. and Uhlenbrook, S. submitted. Lessons learned from catchment scale tracer test experiments using qPCR and natural DNA from total bacteria during rainfall-runoff events in a tropical environment. *Hydrological Processes.*

5.1. Introduction

Quantification of runoff components from different geographical sources during rainfall events is key in catchment hydrology and essential for sustainable management of water resources at catchment scale (Uhlenbrook and Hoeg 2003). Chemical and stable water isotopic tracers are commonly used to decompose the runoff signal of a catchment and determine spatial and temporal sources, respectively. Other researchers have proposed the use of living organisms such as diatoms (e.g. Pfister *et al.* 2009) . The applicability of environmental tracers is limited by its conservativeness, the laboratory analytical technique and the effort for sample collection and the time for sample analysis (Leibundgut et al. 2009). Natural gradient multiple tracer tests are very convenient to gain maximum information of study sites and reduce experimental costs (Ptak et al. 2004). This approach implies the analysis of multiple analites in the lab, which increases time and financial costs. Furthermore, simultaneous use of common tracers such as salt and fluorescent dyes may be limited by lab analysis procedures and local legislations (e.g. radioactive tracers and rhodamine compounds are prohibited in Germany)(Ptak et al. 2004).

In theory, synthetic DNA molecules with individually coded information can be used to perform experiments with unlimited number of tracers (Ptak et al. 2004). Aleström (1995) developed tracing techniques using synthetic DNA molecules with specific coded information. Some advantages of synthetic DNA tracers are its extremely sensitive detection limit with theoretical detection limits of one molecule per sample using quantitative polymerase chain reaction (qPCR) (Watson 1992). Additionally, an unlimited number of DNA sequences can be used which means using different tracers during the same experiment (Ptak et al. 2004). Disadvantages are (unknown) instability in environments with low pH or high microbial activity (Sabir et al. 2000). Also, since DNA molecules are negatively charged, they can be adsorbed by aquifer materials, specially by clay (Ptak et al. 2004). Presence of bivalent magnesium and calcium foster more sorption than monovalent sodium at neutral pH (Lorenz and Wackernagel 1987). Though Aquilanti et al. (2013) reported conservative behavior DNA tracers in a carbonate environment.

Synthetic DNA can be used qualitatively as a tracer in groundwater (Colleuille et al. 1998, Sabir et al. 2000, Sabir et al. 1999). Sabir et al. (1999) worked in a sandy unconfined aquifer in Norway using the forced gradient approach combining single stranded DNA and NaCl. The travel distance for the experiment was 12 m. Using a very dense (5 x 5 m) monitoring piezometer network, they found the DNA tracer at greater depths and distances from the injection point than the NaCl tracer. They explain the result by the high detection limit of the PCR method. A similar experiment using both DNA and NaCl was carried out by Sabir et al. (2000) also in Norway in a fractured rocks environment. The purpose was to investigate the connection between a small lake and a tunnel constructed 180 m below ground. They determined travel times between the lake and the tunnel in the range of 5 to 20 h. This information was used to improve the calibration of a 3D numerical groundwater flow model.

Polymerase chain reaction (PCR) is a powerful molecular detection method, which exponentially amplifies a target gene through thermal cycling by heating and cooling the reaction. However only the final concentration of the amplicon is monitored using a fluorescent dye (Dionisi et al. 2003). In quantitative real time PCR (qPCR) the concentration of the amplicon is monitored using a fluorescent dye allowing the rapid quantification of the amount of template present at the beginning of the amplification process. The fluorescent dye binds with the amplicon during each cycle and emits a fluorescence intensity which reflects the amplicon concentration in real time (Zhang and Fang 2006). The threshold cycle (Ct) is the level at which the reaction reaches a fluorescence above the background value (Heid et al. 1996). The Ct of a sample is inversely proportional to the logarithm of the amount of DNA template in the sample (Harms et al. 2003).

Zhang and Fang (2006) have reviewed environmental applications of qPCR. Key works include Archaea (Suzuki et al. 2000, Takai and Horikoshi 2000, Dionisi et al. 2003), nitrifying bacteria (Hoshino et al. 2001, Okano et al. 2004, He et al. 2007, Li et al. 2011), denitrifying bacteria (Henry et al. 2004), sulfate reducing bacteria (Stubner 2002) and cyanobacteria (Tan et al. 2009).

Some recent attempts to use synthetic DNA as hydrological tracers have been reported mostly for lab or small field scale experiments (100-1,000 m). Ptak et al. (2004) performed lab and field scale experiments in Germany to investigate breakthrough curves and effective transport parameters to delineate water well protection zones. The study site stratigraphy was composed of a top clay layer follow by sand and gravel deposits. Calcium and magnesium concentrations were moderate, pH was neutral and microbial activity was low. Thus, degradation or adsorption of DNA was not expected. At lab scale, undisturbed core samples of 50 cm length and 10 cm diameter were used. Sodium bromide (NaBr) was used as a conservative reference tracer. DNA transport velocities were higher than NaBr, which was explained by size exclusion of the larger DNA molecule and selective flow paths with larger diameter and faster velocities. The field scale experiment (100 m) was performed under forced gradient conditions and sample collection was done at multilevel observation wells. Sodium bromide results showed that breakthrough curves were not always different between the two tracers. However, in some cases DNA did appear first, supporting the results from the lab tests.

Other field scale studies show that synthetic DNA tracers can be used with similar results compared to other chemical tracers such as NaCl, KCl and dyes (Foppen et al. 2011, Aquilanti et al. 2013, Sharma et al. 2012, Foppen et al. 2013). Experiments have been conducted in temperate climates in small (1–2.7 km^2) catchments between Belgium and the Netherlands (Foppen et al. 2013); in a stream (1,000 m) near New York (Sharma et al. 2012), and in karst environments (2,000 m) in Italy. So far, there are no reports of experiments in humid tropical climates.

Recent research on DNA as a tracer has been focused on synthetic DNA, whereas application of natural DNA as a hydrological tracer is not reported in the literature yet. Based on the experiences with synthetic DNA, we tested the applicability of natural DNA (from total bacteria) as a tracer at catchment scale (46 km^2). The use of naturally occurring DNA as a hydrological tracer would avoid the need for tracer injection at multiple runoff source areas. Prior sampling of possible runoff sources would determine suitable end members for hydrograph separation.

The objective of our work is to test the applicability of natural DNA from total bacteria as a tracer to decompose the total discharge signal of a river in a tropical catchment during rainfall–runoff events. Synoptic sampling was performed to determine the background DNA signal.

Although we did not reach these objectives, we present a documentation of our tests and a discussion of shortcomings as well as improvements needed to apply these tracers to hydrological field scale experiments. We believe it is important to share these results with the scientific community since unsuccessful experiments can also lead to instructive lessons, as discussed by Andréassian et al. (2010).

5.2.   Study area

The study area is a 46 km^2 catchment (11 km long) located on the southwestern Pacific Coast of Nicaragua (Fig. 5.1). Soil depth varies between 0.9 m and 0.5 m. The texture of these soils is clay–loam on the surface and deeper in the soil horizon is clay. Land use is dominated by forest (52%), agriculture (20%) and pasture (28%) (UNA 2003). The area is located in the tropical wet forest life zone according to Holdrige's classification (Holdridge 1967). Average year precipitation for this region is 1476 mm year^{-1}. Annual precipitation for 2012 amounted to 1180 mm year^{-1}. Accumulated precipitation during the sampling period (19-25 October 2012) was 275 mm. River discharge during the rainfall-runoff events sampled varied between 0.18 m^3 s^{-1} and 4.68 m^3 s^{-1}.

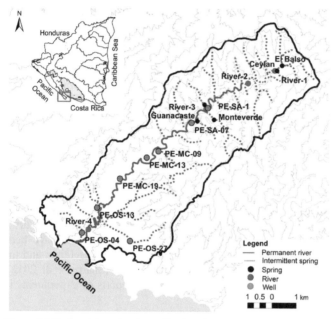

Figure 5.1 Study catchment and sampling locations

5.3.   Methodology

5.3.1. Sampling and nucleic acid extraction

Synoptic sampling of springs, shallow wells and river water was performed in the study catchment once during the dry season (April 2012) and the rainy season (October 2012). Sampling locations are provided in Figure 5.1. Three rainfall–runoff events were sampled every 20 min (Event 1, Event 2 and Event 3). Samples were collected at location River-4 Volume sample was 1,100 ml.

Since samples were collected at a remote location they had to be preserved to be shipped from Nicaragua to the Netherlands. Therefore, samples were filtered *in situ*, first through a 1.5 µm glass fiber filter (Whatman 934–AH) to eliminate impurities and then through a 0.2 µm cellulose acetate filter (Whatman OE66). The latter was stored in an Eppendorf vial with 1.5 ml of pure ethanol. Field blanks (unused filter with ethanol) were also prepared. The vials were refrigerated a 4°C within 12 hours from extraction. They were kept at this temperature until they could be shipped to the Netherlands.

Extraction of NA was performed at UNESCO-IHE qPCR lab. Ethanol was evaporated using a heat block at 65°C for 4 to 5 hours. Once the filters were dry, the method described by Boom et al. (1990, 1999) and also used by Chung et al. (2013) was used to extract nucleic acids (NA) from the filters. NA were eluted in 50 μL of TE–EDTA and then stored at -70°C until qPCR was performed.

5.3.2. qPCR analysis

Samples were analyzed using a quantitative real-time PCR detection system (BioRad MiniOpticon). The primers used for amplification of 16S rDNA gene fragments were 341F, 907RC and 907RA as described by Schäfer and Muyzer (2001).

The qPCR procedure to determine total bacteria DNA consisted in adding 2 μL of DNA extract to 18 μL of qPCR mix containing 10 μL of iQ™ SYBR® Green Supermix from BioRad, 6 μL of DEPC water, 1 μL of primer 341f (5 μM) and 0.5 μL of primers 907rA and 907rc, both at a final concentrations of 2.5 μM.

The efficiency of the qPCR method was assessed through a standard curve. Thereto, we prepared a series of fivefold dilutions (1x to 125x) from NA extracted from a stock *E. coli* solution with an approximated concentration of 10^9 MPN/100 ml. The threshold cycle (Ct) was plotted against the dilution factor to determine the slope of the regression line. Exponential amplification of DNA in PCR is given by Equation 5.1:

$$Rn = Ro\,(1 + E)^n \hspace{3cm} (5.1)$$

where Rn and Ro are the amount of fluorescence signal (proportional to the amount of DNA) after 0 and n cycles, respectively (Schefe et al. 2006) and E is the efficiency of the reaction ($0 \leq E \leq 1$). For an efficiency of 1, the amplicons would double from cycle to cycle. The plot of the log of initial target copy number for a set of standards (series of 3 to 5 orders of magnitude dilutions) versus Ct is a straight line (standard curve). (Higuchi et al. 1993). An efficiency of 1 should yield a slope of 3.3, which is the difference in Ct between each 10 fold dilution.

Samples were diluted 5, 25 and 125 times to reduce inhibition (absence of amplification of the target DNA). Factors inhibiting the amplification of DNA during qPCR act at essential moments during the reaction such as: interference of cell lysis necessary to extract the DNA, degradation or and capture of DNA and inhibition of polymerase activity for amplification of target DNA. Some environmental compounds that act as inhibitors are phenolic compounds, humic acids and heavy metals. Others are non-target DNA and laboratory items such as plastic ware, glove powder and cellulose (Wilson 1997). Humic acids originate from soil they are usually extracted along DNA (Seo et al. 2013).

Since inhibition was still considerable after dilutions, the qPCR mix was modified to include 400 ng μL^{-1} of ultrapure bovine serum albumina (BSA) non-acetylated (Applied Biosystems) in the final volume (Seo et al. 2013, Jiang et al. 2005). The thermal cycling regime consisted of 95°C for 5 min and then 40 cycles of denaturation at 95°C for 30 s and annealing/extension at 55°C for 40 s, 72°C for 40 s and 80 °C for 25 s.

Three non-template controls (NTC) or lab blanks, field blanks duplicates and positive controls (*E. coli*) duplicates were included in each batch. NTCs were prepared by adding DEPC water instead of DNA extract to the qPCR mix.

5.4. Results and discussion

5.4.1. qPCR method

Results indicate that BSA does not affect samples without impurities (*i.e.* NTCs and positive controls) but the effect is higher on natural samples which contain impurities (*i.e.* rainfall–runoff samples). The use of BSA had little influence on Ct values of NTCs and positive controls; with differences of 0.64 and 0.29 in Cts, respectively (Table 5.1). For rainfall–runoff event samples the highest average difference was 1.25 and the highest SD was 1.84. The standard curve yielded a slope of 2.42 (Fig. 5.2) which is equivalent to an efficiency of 73%.

Undiluted samples were not amplified for the most cases. The use of BSA increased the number of amplified samples, especially for Events 1 and 2 (Table 5.2). A combination of 5-fold dilution and the use of BSA allowed amplification of all samples from these two events. However, only 19 out of 24 samples from Event 3 were amplified. For Event 3 it was necessary to perform a 25-fold dilution in combination with the use of BSA in order to achieve amplification of all samples. Three component hydrograph separation based on chloride and $\delta^{18}O$ indicates that Event 3 had the largest surface runoff contribution from all the sampled events (Calderon and Uhlenbrook 2014a). The larger surface runoff contribution explains the need to dilute these samples 25 times in order to eliminate inhibition.

Chung et al. (2013) also found inhibition in environmental water samples composed of a mix of surface water and grey water. In their case, 100- and 1000-fold dilutions were able to exclude inhibition. However, in our case samples contained a large amount of suspended sediments since they are the product of surface and subsurface runoff generation processes (Calderon and Uhlenbrook 2014a). Hata et al. (2014) reported a decrease in recovery efficiency with high levels of suspended solids during rainfall events when studying viruses in river water polluted with sewer overflow. They attribute lower recovery efficiency to membrane pore clogging during sample collection which can be overcome by pre-filtration. As in our case, they also state the possibility of sample loss in the pre-filtration process.

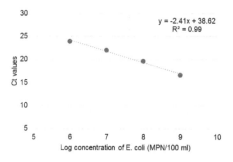

Figure 5.2 Standard curve for total bacteria

Table 5.1 Summary of results for NTCs, field blanks and positive controls

	No samples	Average Ct difference between BSA and non BSA mix	SD
NTC	8	0.64	0.7
Positive control	6	0.29	0.2
Event 1	14	0.22	1.6
Event 2	14	1.18	1.8
Event 3	24	1.25	1.7

Table 5.2 Effect of BSA and dilution factors on rainfall-runoff samples. Surface runoff contribution to total discharge based on 3-component hydrograph separation using chloride and $\delta^{18}O$ as reported in Calderon and Uhlenbrook (2014a)

	No. of samples	No. of samples amplified without dilution	No. of samples amplified with BSA and no dilution	No. of samples amplified with BSA and 5fold dilution	No. of samples amplified with BSA and 25fold dilution	Surface runoff contribution to total discharge (%)
Event 1	11	0	8	11	11	19
Event 2	11	0	7	11	11	32
Event 3	22	3	6	19	22	42

5.4.2. Synoptic samples

DNA from undiluted spring water samples was amplified (Fig. 5.3). The only exception in April was the El Guanacaste spring. DNA from this sample could not be amplified without the use of BSA and a 5-fold dilution. In this case is possible that the site characteristics caused a higher bacterial load or impurities cause inhibition. The Guanacaste spring is located downstream from the other sampling sites and near a small rural community and soils here are deeper (90 cm). The other springs are in more pristine almost inhabited areas. The El Balso sample for October was affected by human error during sample preparation.

The inhibition of qPCR in river water samples from October is likely caused by the influence of surface runoff contribution to river discharge. River water samples collected during the dry season (April) show no inhibition of qPCR, but they do during the rainy season (October), except for River 1. Samples had to be diluted 25-fold and BSA was added to the mix in order to obtain amplification of the sample template.

No inhibition affected qPCR for groundwater samples. This supports the proposition that surface runoff is the source of inhibitory substances in other samples (Figure 5.2). Considering that the losses of DNA in our method are consistent (*i.e.* same loss during NA extraction for all samples and same efficiency of qPCR method) we can qualitatively compare dry and wet season results. We observe lower Cts in April (dry) than in October (wet). This indicates a higher DNA content in dry season samples. Since there was no inhibition in the rainy season samples which may alter the results, we argue that the reduced DNA content is caused by dilution from local recharge from precipitation.

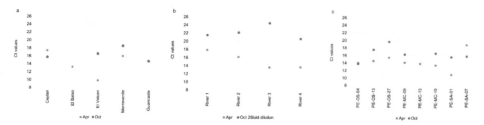

Figure 5.3 Cts for water samples collected during the dry and rainy period. **a** Spring water sample, **b** river water samples, and **c** well water samples

5.4.3. Rainfall-runoff events

We achieved DNA amplification of rainfall-runoff samples using a combination of 25-fold dilutions and BSA. Although no clear Ct trend was found during the hydrograph, we can draw some relevant conclusions from our experiments. We only present results from Event 3 for illustration purposes. The effect of BSA was tested using the same dilution factor (Fig. 5.4). Fourteen out of 21 samples showed lower Cts when BSA was used. The average difference with non BSA samples was 2.1, which is close to the theoretical difference produced by a 5-fold dilution.

The results indicate that sample dilution is not enough to eliminate inhibition in the rainfall–runoff event samples. Addition of BSA to the qPCR mix improves the results. However, further tests are needed to optimize the BSA concentration, since inhibitory substances may change depending on soil characteristics (Wilson 1997, Ptak et al. 2004, Miller et al. 1999). Sample volume also needs to be assessed as Hata et al. (2011) found that large volume samples cause inhibition more frequently than small volumes.

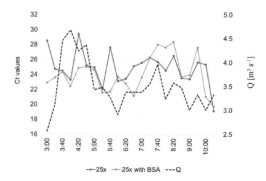

Figure 5.4 Temporal behavior of Cts for diluted samples with and without BSA, from Event 3

5.5. Conclusions

Inhibition of DNA amplification in natural water samples was caused by the influence of surface runoff. This is supported by the differences found in dry and rainy season samples, especially in river water; while groundwater samples were unaffected by inhibitory substances. Bacterial DNA in groundwater samples seems to be diluted during the rainy season, most probably due to the effect of local recharge by precipitation.

The DNA harvesting method used *in situ* allowed sufficient template collection. However, DNA losses during the pre-filtration step have to be evaluated.

Sample dilution and BSA help to reduce inhibition in rainfall–runoff event samples. Nevertheless, additional tests are needed to determine the optimal BSA concentration in the qPCR mix.

Bacterial DNA as a natural tracer is a promising tool for rainfall–runoff process study. However, more field scale experiments are required to use bacterial DNA to investigate these processes in quantitative terms. DNA recovery and qPCR inhibition in runoff samples have to be addressed in future works. Future experiments should include areas with different soil and land use types and different surface runoff generation characteristics.

Chapter 6

Investigation of seasonal river–aquifer interactions in a tropical coastal area controlled by tidal sand ridges

Abstract

Seasonal river–aquifer interactions were investigated in a tropical coastal area where tidal sand ridges control river discharge to the sea. The study site is located in southwestern Nicaragua, dominated by humid tropical hydro-climatic conditions. The aquifer provides water to the rural town of Ostional. Connectivity between the river and the aquifer influences water quality and water availability for humans and for the downstream estuarine ecosystem. The effect of stream stage fluctuations on river–aquifer flows and pressure propagation in the adjacent aquifer was investigated analyzing high temporal resolution hydraulic head data and applying a numerical model (HYDRUS 2D). Tidal sand ridges at the river outlet control the flow direction between the river and the aquifer. Surface water accumulation caused by these features induces aquifer recharge from the river. Simulations show groundwater recharge up to 0.2 m^3 h^{-1} per unit length of river cross section. Rupture of the sand ridges due to overtopping river flows causes a sudden shift in the direction of flow between the river and the aquifer. Groundwater exfiltration reached 0.08 m^3 h^{-1} immediately after the rupture of the sand ridges. Simulated bank storage flows are between 0.004–0.06 m^3 h^{-1}. These estimates are also supported by the narrow hysteresis loops between hydraulic heads and river stage. The aquifer behaves as confined, rapidly transmitting pressure changes caused by the river stage fluctuations. However, the pressure wave is attenuated with increasing distance from the river. Therefore, we concluded that a dynamic pressure wave is the mechanism responsible for the observed aquifer responses. Pressure variation observations and numerical groundwater modeling are useful to examine river–aquifer interactions and should be coupled in the future with chemical data to improve process understanding. The results highlight the importance of preserving the sand ridges to prevent negative alterations to the groundwater–surface water system in this, and other similar catchments in this region.

Based on: Calderon, H. and Uhlenbrook, S. 2014b. Investigation of seasonal river–aquifer interactions in a tropical coastal area controlled by tidal sand ridges. *Hydrology Earth System Science Discussions*, 11(8), 9759-9790.

6.1. Introduction

Groundwater and surface water are both intrinsically related components of the hydrological cycle (Winter 1998). The exchange of water between streams and aquifers influences the quality and quantity of water within both domains according to the fluxes and water chemistry of the water moving through the streambed and the changes that occur at the groundwater–surface water interface (Blöschl 2006). These interactions are essential for water supply, water quality and aquatic ecosystems (Sophocleous 2002, Alley et al. 2006) and thus they are important for the sustainable management of water resources.

The movement of water between groundwater and surface water provides a major pathway for chemical transfer between terrestrial and aquatic systems (Winter 1998). Hydraulic connections between groundwater and surface water also provide a conduit for the potential transport of contaminants (Oxtobee and Novakowski 2002, Savenije 2009, Mendoza et al. 2008, Chaves et al. 2008). These flows are controlled by the magnitude and distribution of hydraulic conductivities, the relation of stream stage to the adjacent water table and the geometry and position of the stream within the alluvial plain (Woessner 2000). Kalbus et al. (2006) provides an extensive review of different methods to investigate groundwater-surface water interactions and states the need of a multi-scale approach which combines different methods to help quantify the fluxes estimates between rivers and aquifers. Temperature differences between surface water and groundwater are often used in combination with hydraulic gradients to investigate these interactions (Silliman and Booth 1993, Conant 2004, Anderson 2005, Krause et al. 2012, Bartsch et al. 2014, Westhoff 2007).

Piezometer transects across rivers are commonly used to monitor hydraulic gradients and investigate river–aquifer fluxes (Conant 2004, Mendoza et al. 2008, Bartsch et al. 2014). However, hydraulic gradients between the river and the aquifer indicate only pressure distributions and quantification of fluxes requires knowledge of sediments hydraulic conductivities (Kaser et al. 2009). In lowland rivers, complex heterogeneities of the sediment and therefore the hydraulic conditions, pose a major challenge to use this approach (Krause et al. 2012).

Individual precipitation events may change the direction of water exchange between the surface and subsurface on a daily or even shorter time horizon, due to localized recharge near the stream banks, flood peaks moving downstream or transpiration by river bank vegetation (Alley et al. 2006). Highly variable exchange fluxes are reported for intense precipitation seasons (Mendoza et al. 2008, Bartsch et al. 2014).

Bank storage occurs when stream water infiltrates in the adjacent aquifer during the rising stage of a flood. During the recession stage, the stored water moves back into the stream. Bank storage may attenuate the flood peak (Pinder and Sauer 1971) and increase the base time of the hydrograph (Chen and Chen 2003). Release of stored water into the stream may also alter stream water quality (Squillace 1996). Water table fluctuations may also occur through pressure exchange between surface water and groundwater without flow mixing (Wondzell and Gooseff 2013). Pressure wave propagation in floodplains has been analyzed by several authors using analytical solutions (Welch et al. 2013), principal component analysis of hydraulic head data (Lewandowski et al. 2009), cross-correlation of hydraulic head data, (Jung et al. 2004, Vidon 2012, Cloutier et al. 2014), numerical modeling (Sophocleous 1991, Chen and Chen 2003) and combination of analytical and numerical methods (Barlow et al. 2000).

The research site is located in the flat coastal catchment area of a tropical river; the surrounding aquifer provides water to the small rural town of Ostional, Nicaragua. Connectivity between the river and the aquifer influences water quality and water availability for humans and for the downstream estuarine ecosystem (Calderon et al. 2014). Therefore, understanding the local recharge mechanism and discharge to the estuarine ecosystem are very important for water management of this area.

Examination of the interaction mechanisms between groundwater and surface water in tropical regions may differ from observations in temperate regions, based on the difference in rainfall intensity and seasonality (Bonell 1993). Scarcity of detailed investigation in tropical regions leaves a gap in the scientific understanding of groundwater–surface water interactions. Transfer of commonly used hydrological methods from one hydro-climatic region may pose unexpected challenges. Therefore, our study is addressing the knowledge gap regarding river-aquifer interactions under tropical rainfall–runoff conditions.

The objective of this work was to investigate the stream–aquifer interaction mechanisms and flow rates in a system where tidal sand ridges control the river discharge to the sea. The variability of the system during wet and dry conditions is also studied by: i) analysing of hysteretic patterns of river stage and hydraulic heads, and ii) 2D numerical modeling simulating pressure propagation in the groundwater system during river stage changes.

6.2. Study area

The study was carried out at the flat lower part of a coastal catchment located in the southwestern Pacific of Nicaragua. An experimental cross section (Fig. 6.1) at the Ostional River was investigated for the period of March 2012–April 2013. The cross section is located next to the town with the same name. Population throughout the catchment is about 1500 people, most of which lives near the coast. Land use in the catchment is forest, subsistence agriculture and grazing. Agricultural is the main land use around the experimental site. Some patches of forest can also be found.

Local geology is composed of a 10–15 m thick clay–alluvial deposits unit on top of a fractured shale unit of unknown thickness. The top clay layer is discontinuous and relatively narrow (5 m). Hydraulic conductivity estimates for the upper unit range between 0.33 m d^{-1} and 6.7 m d^{-1} and the estimated value for the shale unit is 9.07 m d^{-1} (Calderon et al., 2014).

River channel geometry at the site is 18 m wide and streambed is composed of clay and silt. River banks are 1.8 m high on the West side and 1 m high on the East side. Both banks show a mixed matrix of alluvial materials and clay. River stage fluctuated between 0.15 m and 0.85 m during the study period. River width varied between 1 m to bankfull, depending on precipitation events.

The geomorphology of coast in the study area is determined by strong waves and littoral drift, which produces sand ridges along the coast line. These geomorphologic features control freshwater and seawater mixing and in/outflows of rivers and, consequently, the abiotic conditions of coastal ecosystems such as the mangrove forest just downstream of the study area (Calderon et al. 2014). Sand ridges are common on the Pacific Coast of Central America (Jimenez et al. 1999). Surface water accumulates in the estuary until the water levels overtop the ridges and rupture the ridges (Calderon et al., 2014).

The rainy season spans from May to November. October is the rainiest month with an average accumulated precipitation during the study period of 618 mm month^{-1} and a maximum of 1016 mm month^{-1}.

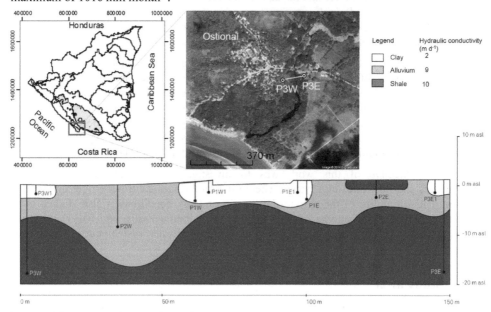

Figure 6.1 Stratigraphic interpretation at the piezometric cross section based on macro- and micro-analysis of sediment samples (Calderon et al. 2014). Piezometer set up across the river; West (W) and East (E) piezometers, number indicates position with respect to the river banks, 1 is the closest 3 is the farthest from the river. Secondary numbering indicates same location but different depth (i.e. P1E and P1E-1)

6.3.   Methodology

6.3.1. Piezometric cross section

Ten piezometers were installed at a cross section of the river (Fig. 6.1). Distance between piezometers is 25 m. Piezometer depth increase away from the river on the East and West side. Shallowest piezometers are closest to the river banks. Drilling samples were correlated to produce a stratigraphic model (Calderon et al. 2014). Piezometers were completed at the end of February 2012 and divers were installed in early March 2012 and remained installed until April 2013.

Water table fluctuations and temperature were monitored continuously every 30 minutes in every piezometer and at the river, using Schlumberger mini divers, range 10 m, DI501 (accuracy ±0.005 m and ±0.1°C). Barometric compensation of the data was performed using atmospheric pressure data recorded at the same interval and for the same time period, using a mini baro diver, range 1.5 m (Schlumberger 50013, DI501).

Elevations at the cross section were determined using a differential GPS with an average error of ±0.009 m. An electrical resistivity tomography (ERT) survey was performed at the cross section. The ABEM Lund Imaging System (Dahlin 1996) was used with a Schlumberger array with a spacing of 5 m. The ERT profile was 300 m long and reached a depth of 60 m. Data was inverted using RES2DINV (Loke and Barker 2004). Resistivity ranges for each stratigraphic unit were identified through correlation with drilling samples. The ERT profile was then used to delineate the stratigraphy.

6.3.2. Numerical model

Pressure head changes at the piezometric cross section were simulated using HYDRUS 2D (version 1.08). This finite element software numerically solves Richard's equation for variably saturated water flow and the convection–dispersion equation for heat and solute transport (Šimůnek et al. 2012). We decided to model only pressure heads since the temperature fluctuations in our study area are very small (less than 1°C) to reliably reproduce them with limited knowledge of heat transport properties of the streambed and aquifer materials.

The model domain is 160 wide and 30 m deep. The triangular mesh was automatically generated with an optimal size of 0.6 m. The mesh was refined to 0.2 m at the boundaries of the model and around pressure head observation points (piezometers). The model time step was 30 min for the longer simulation periods and 0.2 min for the short simulation periods.

Boundary conditions were defined as variable pressure head at the West and East limits and the river bed. The top of the model is defined as an atmospheric boundary on the East side, which allows evaporation and infiltration. On the west side, the top of the model was defined as a variable seepage to atmospheric boundary condition. The seepage boundary condition allows groundwater flow when the water table rises near the model surface. When the water table is below the surface the boundary condition changes to atmospheric, allowing infiltration and evaporation.

We assumed that there is no significant downward groundwater flow below the deepest piezometers, since the observed hydraulic gradient is upwards. Thus, the bottom of the model is defined as a no flow boundary. Initial conditions were defined by the observed pressure heads.

The stratigraphic model consists of three materials: clay, alluvium and shale (Fig. 6.1). The van Genuchten-Mualem soil hydraulic model was used to describe soil hydraulic parameters. Initial values of saturated hydraulic conductivity (Ks) for each material were defined based on slug test results (Calderon et al. 2014). Parameters of the van Genuchten-Mualem model were estimated using soil textures and the computer program Rosetta (Schaap et al. 2001) which uses pedotransfer functions to predict van Genuchten's water retention parameters and saturated hydraulic conductivities. The selected soil textures are based on our initial estimates of Ks. Rainfall and evaporation inputs were based on data from a monitoring station in the study area (Calderon et al. 2014).

Since our model is 2D, calculated flows are given in $L^2 T^{-1}$. Therefore, flows were multiplied per unit length (1 m) of the river cross section in order con convert them to $L^{-3} T^{-1}$ units.

6.3.3. Model calibration

HYDRUS implements the Marquardt-Levenberg parameter estimation technique. The method combines Newton's and steepest descend methods to generate confidence intervals for K_s (Šimůnek et al. 2012). The optimization process is based on the minimization of the objective function which states the difference between observed and estimated values. The objective function defined by Šimůnek et al. (1998) accounts for deviations between measured and calculated space-time pressure heads and differences between measured and calculated Ks (Šimůnek et al. 2012). The objective function is defined as:

$$\phi(b,p,q) = \sum_{j=1}^{mq} v_j \sum_{i=1}^{nqj} w_{i,j} \left[q_j^*(x, t_i - q_j(x, t_i, b)\right]^2 + \sum_{j=1}^{mp} \bar{v}_j \sum_{i=1}^{npj} \bar{w}_{i,j} \left[p_j^*(\theta_i) - p_j * (\theta_i, b)\right]^2 +$$
$$\sum_{j=1}^{nb} \hat{v}_j \left[b_j^* - b_j\right]^2 \tag{6.1}$$

where the first term on the right hand side represents the deviation between measured and calculated space-time variables, m_q is the number of different sets of measurements and n_{pj} is the number of measurements in a particular set. q^*_i represents specific measurements at time t_i for the j_{th} measurement at a location $x(r, t)$, $q_j(x, t_i, b)$ are the model predictions for the vector of optimized parameters b, v_i and $w_{i,j}$ are the weights associated with a particular measurement set or point, respectively. The second term is the difference between independently measured and predicted soil hydraulic properties. The terms m_p, n_{pj}, $p^*_j(\theta_i)$, $p_j(\theta_{i,b})$, v_j and $w_{i,j}$ have similar meaning than in the first term, but for the soil hydraulic properties. The last term is a penalty function for deviations between prior knowledge of the soil hydraulic parameters b^*_j and their final estimates b_j, n_b is the number of parameters with prior knowledge and $\hat{v}_j$ represents pre-assigned weights.

The model was calibrated for the dry period of 1 December 2012–10 March 2013 by inversely calculating Ks. The Ks values of the three materials were iteratively adjusted with the objective function until the simulated pressure heads approximated the observed values. The range of Ks values was constrained by the estimations from slug tests. A set of 506 pressure head (h_p) observations were used to minimize the objective function to 0.012. The R^2 between predicted and observed values was 0.99. Optimization parameters are presented in Table 6.1 along with soil water retention parameters estimated by the Rosetta software. The same model was used to simulate a second set of pressure head observations from the rainy period of 1 October 2012–25 October 2012.

Table 6.1 Soil hydraulic parameters and optimized K_s values (m d^{-1}) for each material

Material	θs (-)	θr (-)	α (m^{-1})	n (-)	Initial value	Min	Max	Optimized value
	\multicolumn{4}{Estimated by Rosetta}	\multicolumn{4}{Optimized Ks values by inverse solution (m d^{-1})}			
Clay	0.3	0.1	5.9	1.48	0.3	0.3	2	2
Alluvium	2.6	0.04	14.5	2.68	6	5.0	12	9
Shale	4.1	0.05	3.31	4.15	9	5.0	13	10

6.3.4. River stage change simulations

Changes in river stage were simulated for the rainiest period within the time span of the study, from 1 October 2012 to 30 October 2012, in order to look into short-term changes in the river-aquifer interactions. Additionally, the effect of larger river stage fluctuations were investigated through modeling. Since 2012 was a relatively dry year compared to 2011 and 2010 (Calderon and Uhlenbrook 2014a) no bankfull events were observed. Thus, a hypothetical bankfull event was recreated based on the highest precipitation records between 2010 and 2012. We selected a rainfall event from 19 October 2011 which had an accumulated precipitation of 100 mm during 5 hours. Unfortunately, no river stage data was available before 2012. Therefore, river stage was inversely estimated using a rating curve from a rainfall event occurred in 2012 with an R^2 of 0.84 (n=21).

River discharge was estimated assuming that all precipitation in excess of soil infiltration capacity generates surface runoff (Fetter 2001). We did not consider depression storage. The contributing area was assumed to be the alluvial valley of 1 km^2, since for the simulated event no precipitation was observed in the upstream catchment area (Calderon and

Uhlenbrook 2014a). Infiltration capacity was estimated in 10 mm hr^{-1} (Calderon and Uhlenbrook 2014a). Two sets of initial conditions were used for the simulation of the hypothetical event: dry and wet period. Thus, the influence of the initial water table could be also analyzed. They were defined based on the hydraulic pressure distribution results from the calibration period.

6.4. Results

6.4.1. Hydraulic head fluctuations and sand ridges

River stage and hydraulic head fluctuations (Fig. 6.2) show a synchronous pattern throughout the study period, indicating a strong hydraulic connection between the river and the adjacent aquifer. During the first two months of monitoring, the river course was temporarily deviated to construct a bridge. This caused the irregular behavior observed during April–June 2012. Daily precipitation, river stage and piezometer hydraulic heads are presented in Figure 6.2. Daily precipitation (Fig. 6.2a) was usually below 25 mm d^{-1}, except in October when the highest value reached 140 mm d^{-1} (25 October).

A sustained increase in hydraulic heads was observed from April to June 2012, despite the lack of precipitation. The river stage also increased during this period. The increases in both surface and groundwater levels are explained by groundwater discharge coming from the upper catchment. The presence of sand ridges prevents surface water discharge to the ocean, causing surface water accumulation and aquifer recharge from the river. Sudden drops in river stage were caused by rupture of the beach ridges (Calderon et al. 2014). Small precipitation events caused peaks in river stage and groundwater levels (*i.e.* end of June 2012).

Two large river stage increases (0.6 m each) and a smaller one (0.3 m) were observed during October 2012 (Fig. 6.2c). The first occurred between the 1 October and 6 of October. However, the rising part of this peak is missing due to technical problems. The second peak occurred between the 14 October and 20 October. The third peak occurred between the 21 October and 31 October. River bank piezometers experienced a hydraulic head increase equal to the increase in river stage for peaks 1 and 2. Piezometers located farther from the river bank suffered an increase of 0.5 m. In the case of peak 3, hydraulic heads on the west side of the river increased 0.2 m and only 0.1 m on the east side.

Local precipitation in Ostional was exceptionally high, 700 mm month^{-1} for October 2012. Whereas upstream rain stations recorded 263 mm month^{-1} and 289 mm month^{-1} (Calderon and Uhlenbrook, in review). Accumulated precipitation from 6 October to 19 October was 322 mm in Ostional, which caused the second river stage peak. However, the amplitude of this peak was also enhanced by the beach sand ridges blocking the river outlet to the ocean. The precipitation event during the night of 19 October 2012 (134 mm d^{-1}) caused excess of stored water which induced the ruptured of the beach ridges. The system released the stored water on the 20 October between 9:00 and 17:00. The third peak was caused by the accumulated precipitation between the 20 October and 26 October which amounted to 361 mm.

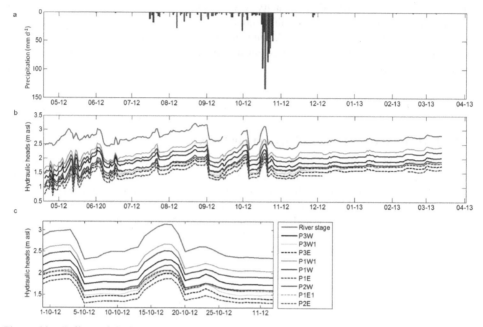

Figure 6.2 a Daily precipitation, **b** mean daily river stage and hydraulic head fluctuations for the period of April 2012 to March 2013, and **c** zoom of river stage and hydraulic head daily average fluctuations for October 2012; locations are given in Figure 6.1

6.4.2. Surface water and groundwater temperature fluctuations

As the river temperature dropped in early June due to precipitation, so did the temperatures in the shallow piezometers. Groundwater temperatures for the shallow piezometers were between 28°C and 31°C between April and June 2012 (dry season). On the East side, groundwater temperature stayed stable at around 28°C. Groundwater temperature on the West side gradually decreased between mid-June and mid October 2012 (rainy period) to values between 28°C and 28.5°C. The deepest piezometers located farthest from the river (P3W and P3E) show stable temperatures. P3E shows temperature at around 29°C. Only during the peak of the rainy season a drop of 0.3°C was observed. Highest temperature in P3W was 29.5°C, in mid-June 2012. During mid-October 2012 the temperature slowly decreased to 28.6°C and stayed stable until the end of the study.

6.4.3. Hysteretic patterns in hydraulic heads

Hysteresis patterns were analyzed for the periods with and without the presence of sand ridges. Daily average hydraulic heads were plotted against river stage (Fig. 6.3) for the second and third river stage peaks. In the case of the second peak all hysteretic patterns are counterclockwise. The counterclockwise pattern indicates lower hydraulic heads during the rising limb and higher heads during the falling limb. Thus, demonstrating bank storage. However, the hysteresis loops are narrow. This indicates that the changes in groundwater levels between the rising and falling limb of the hydrograph are small, only about 0.1 m. For the third peak the loops are not completely closed because the river stage and hydraulic heads were higher at the beginning of this period due to the influence of the second peak.

During both periods we observed higher hydraulic heads on the west side of the river. Also the heads in piezometers located closest to the river indicate a downward gradient from the shallowest piezometers (P1W1 and P1E1) towards the deeper piezometers (P1W and P1E). However, an upward gradient is observed between the piezometers located farthest from the river (P3W-P3W1 and P3E-P3E1). However, we do not show results from P3E1 because during the rainy period this piezometer reacted immediately to precipitation events, probably because of precipitation shortcutting around the base of the piezometer.

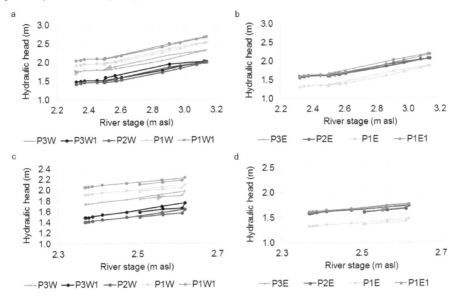

Figure 6.3 Daily averages of river stage and hydraulic head hysteresis plots: **a** and **b** correspond to the period from 6 October 2012 to 20 October 2012 (with the presence of sand ridges); **c** and **d** correspond to the period from 21 Oct to 31 October 2012 (without the presence of sand ridges)

6.4.4. Statistical evaluation of model calibration

Statistical analysis of the model performance reflects the fact that the model is capable of reproducing the pressure head dynamics for both periods. Correlation coefficients (R^2) vary between 0.88 and 0.99. Highest root mean square error (RMSE) is 0.3 for the dry period and 1.0638 for the rainy period (Table 6.2). Mean absolute error (MAE) varies between 0.07 m and 0.22 m for the dry period; and between 0.06 m and 1.3 m for the rainy period. The MAE indicates the difference between observed and simulated pressure heads.

Table 6.2 Statistical analysis of model performance

Observation points	Dry period			Rainy period		
	MAE (m)	R^2 (-)	RMSE (-)	MAE (m)	R^2 (-)	RMSE (-)
P3W	0.0816	0.97	0.0822	0.3183	0.99	0.3095
P3W1	0.2207	0.62	0.2307	0.8647	0.98	0.9398
P2W	0.4679	0.92	0.4682	1.0328	0.92	1.0638
P1W	0.1597	0.98	0.1598	0.4849	0.95	0.4830
P1W1	0.1875	0.99	0.1876	0.3023	0.95	0.3120
P1E1	0.2899	0.94	0.3075	0.9170	0.91	0.9014
P1E	0.1708	0.89	0.6218	1.2933	0.91	1.2707
P2E	0.1880	0.93	0.2194	1.2346·	0.88	1.1916
P3E	0.0691	0.99	0.0692	0.0572	0.95	0.0776

6.4.5. Simulation of river stage changes

6.4.5.1. Rainy period

The rainy period of October 2012 was simulated. Pressure head dynamics in all piezometers were well simulated, except in P3W1 and P3E, probably because of local heterogeneities not captured by the model. Results for the shallowest piezometers are shown in Figure 6.4. Model performance on the west side was better. Average of the MAE for west piezometers was 0.55 m, and for East piezometers was 0.66 m.

The synchronous response in all piezometers to river stage changes was examined by looking at groundwater flow velocities between the river and the piezometer locations. Average linear velocity estimated for river bank piezometers assuming an effective porosity of 0.3 (-) yields values of 3 m d^{-1}; for the other piezometers estimates are between 0.3 m d^{-1} and 0.4 m d^{-1}. The numerical model estimated a maximum groundwater effective velocity of approximately 0.3 m d^{-1}. These velocities cannot explain the timing of the water table response. It would take between 7 hours (for riverbank piezometers) and 100 days (for the farthest piezometers), for groundwater recharge from the river to reach them mainly due to the low slope and the given hydraulic conductivities. Therefore, we conclude that the change in pressure heads is driven by a nearly instantaneous pressure wave from the river towards the aquifer. This wave is, however, attenuated by up to 50% when it reaches the piezometers located farthest from the river (horizontal distance 50 m).

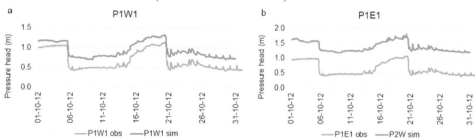

Figure 6.4 Observed and simulated pressure heads every 30 min for shallowest piezometers for the period 1-31 October 2012

The peak groundwater recharge was similar for 6 October and 20 October, approximately 0.08 m³ 30 min⁻¹ (Fig. 6.5). In both cases river stage increase was 0.6 m and the sudden change in flow direction was caused by rupture of sand ridges. Maximum groundwater exfiltration occurred immediately after the rupture of sand ridges and it reached 0.04 m³ 30 min⁻¹. Beginning on 21 October groundwater recharge has a different dynamics. Recharge reached approximately 0.02 m³ 30 min⁻¹. The change in flow direction was gradual. This is because the river was flowing freely into the ocean after the rupture of sand ridges.

After 26 October, without sand ridges present and without precipitation, groundwater exfiltration reached approximately 0.02 m³ 30 min⁻¹. Nevertheless, it remained near that value for about 5 days (Fig. 6.5). The difference in the duration and magnitude of groundwater exfiltration with and without sand ridges present, indicate the large control of these geomorphologic features on groundwater–surface water interactions.

Small periodic shifts are observed in the flow direction. They have a periodicity of about 24 hours and given the lack of precipitation, they are attributed to high tide influence on the river stage. Mixed semidiurnal tides were observed during this period, with two high tides of different amplitude during 24 hours. However, only the larger amplitude tide had an effect on the streambed flow.

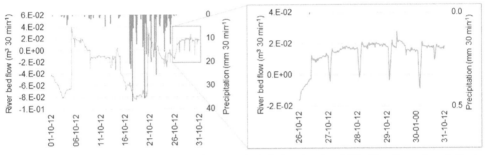

Figure 6.5 Precipitation and flow across the streambed during October 2012. Negative values indicate groundwater recharge and positive values indicate groundwater exfiltration

6.4.5.2. Hypothetical river stage peak

For dry period initial conditions, groundwater recharge occurs during the entire simulation period (Fig. 6.6a). Groundwater recharge is higher during peak river stage (at 210 min) and reaches 0.1 m³ 30 min⁻¹. Analysis of flows across the river banks shows groundwater exfiltration after peak river discharge (Fig. 6.6b). The West bank continues to exfiltrate groundwater after the peak, reaching 0.002 m³ 30 min⁻¹. The East bank exfiltrated groundwater only during peak river stage, reaching 0.005 m³ 30 min⁻¹.

Wet initial conditions produce a different pattern in flow direction. Groundwater recharge occurs only during peak river stage and reaches 0.05 m³ 30 min⁻¹ (Fig. 6.6c). Groundwater exfiltration starts before the river stage peak. Exfiltration is caused by an increase in groundwater levels due to recharge by precipitation. Exfiltration on the West bank reached 0.007 m³ 30 min⁻¹ and on the East bank it reached 0.026 m³ 30 min⁻¹ (Fig. 6.6d).

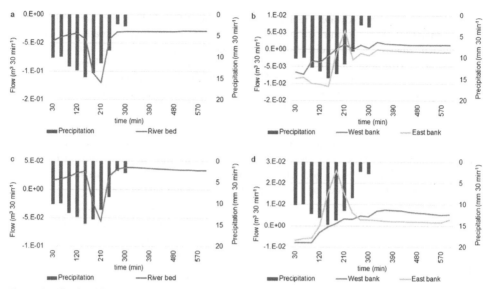

Figure 6.6 Simulated flow rates for a hypothetical stream stage peak: **a** and **b** dry initial conditions; **c** and **d** wet initial conditions

6.5. Discussion

6.5.1. Effect of sand ridges on groundwater–surface water flows

Surface water accumulation due to the presence of sand ridges in the river mouth controls the flow direction between the river and the aquifer (Fig. 6.7). The observed fluctuations in river stage and hydraulic heads and the river–aquifer flows estimated by the model, indicate that the presence of sand ridges causes groundwater recharge. The largest observed river stage increments (0.6 m) caused an approximate increase of 0.5 m in hydraulic heads. Simulations also indicate groundwater recharge up to 0.08 m^3 30 min^{-1} from the stream during these periods. The rupture of sand ridges on 20 October 2012 caused a sudden release of surface water into the ocean. This release also caused a reversal in the direction of the river–aquifer flow direction, allowing groundwater exfiltration of up to 0.04 m^3 30 min^{-1}.

6.5.2. Bank storage

Hysteresis loops of river–stage and hydraulic heads show a counterclockwise pattern, indicating bank storage. However, the loops are narrow which indicates that relative changes in hydraulic head between the rising and falling limbs are small. Hydraulic head response to river stage changes is fast. In the case of the second peak in river stage, heads returned to their initial values 10 hours after the rupture of the sand ridges. A fast response is also observed for individual events occurred after the rupture of sand ridges. The first event had a river stage peak of 0.3 m and stored water was released within 10 hours after the peak. The second event had a river stage peak of 0.16 m and stored water was released within 5 hours after the peak.

6.5.3. Effects of river stage changes on groundwater recharge

Simulations indicate that most stream infiltration occurs through the river bed by vertical flow. Hydraulic gradients across river banks shift with peak river stage. The effect is shorter in duration in the East river bank. The flow across the river banks are small (0.04 m³ 30 min⁻¹ to 0.2 m³ 30 min⁻¹). This result is supported by the hysteresis loops, which show small changes in groundwater levels. The amount of stored water is small and it is released within a short time period (10 hours). Bartsch et al. (2014) found frequent river-aquifer flow reversals during monsoon events in South Korea. They explained that the changes in hydraulic gradients are caused by intense precipitation events which caused increases in river stage. In our case river stage increases are controlled by sand ridges.

Observed and simulated pressure heads show a synchronous behavior with river stage. This response cannot be explained by Darcian flow. Average linear groundwater flow velocities between the river and piezometers are estimated between 0.3 m d⁻¹ and 3 m d⁻¹. The estimated times for a flood wave to arrive to the piezometers nearest to the river is 7 hours and to the piezometers farthest away from the river is 100 days. Therefore, we infer that the changes in pressure heads are driven by a nearly instantaneous dynamic pressure wave from the river towards the aquifer. This wave is attenuated by up to 50% when it reaches the piezometers located farthest from the river. The fast propagation of the pressure wave may be explained by the clay unit in the aquifer. Sophocleous (1991) found that the Grand Bend Prairie aquifer in Kansas, behaved as confined because of the widespread shallow and thin clay layers within the aquifer. The aquifer had low storativity but high transmissivity. Thus, pressure waves from the streams travel fast and to great distances (tens of kilometers). The pressure waved caused by stream stage changes was explained as fluid molecules transmitting pressure between each other, with very small spatial displacement (Sophocleous 1991).

Following this line, Jung et al. (2004) explained the water table response in a floodplain to river stage increases through the kinematic wave mechanism, since the synchronous response could not be explain by the much lower Darcian velocities. Cloutier et al. (2014) also found that Darcian flow was not sufficient to cause flood wave propagation from a river into the adjacent aquifer in the floodplain. Thus, they proposed a dynamic wave mechanism as explanation for the water table response since the hydraulic head fluctuations were not conservative in time and space. The difference between both mechanisms is that the kinematic wave is non-dispersive and non-diffusive, whereas the dynamic wave is dispersive and diffusive (Cloutier et al. 2014). Vidon (2012) found that Darcy´s velocities were 2 to 3 order of magnitude too small to account for the water table response to river stage increases. Also, they showed that infiltration from precipitation was not responsible for the fast water table response, since electrical conductivity of riparian groundwater remained stable instead of decreasing. Therefore, they explained the rapid water table rise by means of a kinematic wave processes. Lewandowski et al. (2009) also found a dynamic wave mechanism in the pressure propagation between a stream and an aquifer. They found that Darcy´s velocities were 100 times slower that the pressure wave propagation in an alluvial unconfined aquifer. They also found dampening of the pressure wave with distance from the river. In our case, the attenuation of the pressure wave with distance from the river indicates that the mechanism for propagation is a dynamic wave.

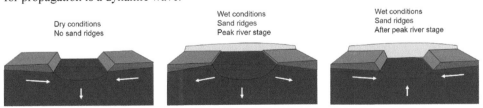

Figure 6.7 Conceptualization of sand ridges influence on river– aquifer flows

6.6. Conclusions

Tidal sand ridges at the river outlet control the flow direction between the river and the aquifer. Surface water accumulation caused by these features at the river outlet induces aquifer recharge from the river. The simulations show that the larger river stage increases caused by sand ridges increase groundwater recharge significantly. Rupture of the sand ridges (by overtopping and erosion) causes release of the stored surface water and a sudden shift in the direction of flow between the river and the aquifer. The difference in the duration and magnitude of groundwater exfiltration with and without sand ridges present, indicate the large control of these geomorphologic features on ground water–surface water flow exchange.

Bank storage occurs during stream stage increases. However, the volume of stored water is small (0.004 m^3 h^{-1} to 0.06 m^3 h^{-1}). Stored water is quickly released back into the stream after the peak river stage. The hydraulic gradients across the river banks show shifts during flood events. During dry conditions a flood event causes a reversal in the hydraulic gradient. After peak river stage, river banks release water into the stream, although in very small amounts. During wet conditions, river bank flow becomes more significant. However, under both scenarios, vertical streambed infiltration remains the most important component of aquifer recharge during stream stage peaks.

The synchronous response of the water table to river stage fluctuations cannot be explained solely based on linear average velocities. Estimated travel times from the river to the observation points are much longer than those observed. The nearly instantaneous response of the aquifer to the pressure wave caused by the stream stage increased is explained by the presence of the discontinuous clay layer on top of the aquifer.

The aquifer behaves as confined, rapidly transmitting pressure changes caused by the river stage increase. The pressure wave is attenuated with increasing distance from the river. Therefore, we propose a dynamic pressure wave as the mechanism responsible for the observed aquifer response; as opposed to the kinematic wave mechanism, which would not be attenuated with time nor distance.

The results highlight the need to preserve the sand ridges in this, and other similar catchments of the region; in order to prevent negative alterations to the groundwater–surface water system. This consideration needs to be included in water resources management plans.

Chapter 7

Conclusions and recommendations for future research

7.1. Conclusions

7.1.1. Geomorphological controls

Catchment structure (defined by topography, geology and land use) controls surface and subsurface runoff generation. The stratigraphic and topographic characteristics of the catchment determine two major groundwater flow systems (Chapter 2): one regional system located in the shale/limestone unit and one local located in the clay/alluvium unit. The first one is recharged upstream in the catchment and it is characterized by high chloride and silica concentrations and heavier water isotope content. The second one is recharged locally by precipitation and river infiltration. It is characterized by low chloride and silica concentrations and lighter water isotope content. The regional system sustains baseflow during dry periods.

Another key structural characteristic is the tidal sand ridges, which determine the water balance for the mangrove ecosystem at the catchment outlet (Chapter 3). Sand ridges prevent river discharge into the ocean, inducing surface water accumulation in the lower catchment area. This specific characteristic of the catchment causes: i) positive water balance which enables the subsistence of the mangrove forest during dry periods; and ii) induces aquifer recharge by increasing river stage during dry and wet periods (Chapter 6). This is very important for the alluvial aquifer which provides water for the town.

Surface runoff generation is controlled also by other structural features of the catchment defined by the scale, stratigraphy, slopes, soil and land use (Chapter 4). Steep slopes, forested hillslopes, permeable soils and less permeable shale layers favor subsurface stormflow in the upstream sub-catchments. At the catchment level, a broader valley, thicker alluvial deposits and smooth slope favor major contribution of groundwater.

Tidal sand ridges at the river outlet control the flow direction between the river and the aquifer. Numerical simulations show that the larger river stage increases caused by sand ridges increase groundwater recharge significantly. Rupture of the sand ridges causes release of the stored surface water and a sudden shift in the direction of flow between the river and the aquifer. The difference in the duration and magnitude of groundwater exfiltration with and without sand ridges present, indicate the large control of these geomorphologic features on groundwater–surface water flow exchange.

Aquifer response to increases in river stage are controlled by the presence of a thin (<3 m) discontinuous clay layer, which causes the quasi-instantaneous pressure head changes in the aquifer in response to increases in river stage. The aquifer behaves as confined, rapidly transmitting pressure changes caused by the river stage increase. The pressure wave is attenuated with increasing distance from the river, in the form of a dynamic pressure wave.

7.1.2. Hydro-climatic controls

Highly temporal and spatial variability of precipitation, even for a relatively small catchment, affects availability of water resources for specific ecosystems (Chapter 3) and humans (Chapter 4), determines sources of surface runoff generation (Chapter 4) and induces changes in groundwater–surface interactions (Chapter 6).

River baseflow is sustained by groundwater during the dry season. Groundwater recharge occurs during the rainy season in the upper catchment area. Secondary porosity in the shale unit permits recharge. Groundwater flows towards the lower catchment area through the shale unit. Precipitation recharge induces piston flow, pushing previously recharged groundwater (Chapter 2). Groundwater recharge from exceptionally wet years (*i.e.* 2011) compensates for relatively dry ones (*e.g.* 2012), thus maintaining a positive net water balance (Chapter 4). Antecedent precipitation conditions determine the occurrence of significant surface runoff component in the hydrograph at catchment scale. River stage increases due to precipitation events induce aquifer recharge in the lower catchment area (Chapter 6).

7.1.3. Water resources management

There are key areas within the catchment which control groundwater recharge and runoff generation. These are the upstream forested sub-catchments (Chapters 2 and 4) and the tidal sand ridges (Chapters 3 and 6). Forested hillslopes in the upper catchment area are crucial to reduce surface runoff and therefore soil erosion. Groundwater recharge occurs through the fracture network in the shale/limestone in this area and travels downstream the catchment via subsurface flow paths. Although currently water supply needs are satisfied by shallow private wells, future touristic development of this coastal area supposes further exploitation of groundwater resources. The few existing deep wells are located in the shale unit, which receives groundwater recharged mostly in the upper in the catchment. Land use changes in the currently forest areas may induce drastic changes in the hydrologic regime of the catchment and the quality of the water resources. Soil erosion by increased surface runoff is another threat with multiple impacts for water resources, water quality and ecosystem health.

In the lower catchment area, tidal sand ridges control surface water-groundwater exchange flows and also guarantee preservation of the mangrove ecosystem during dry periods. However, flooding of this area during the rainy season prompts the local community to remove the sand ridges. So far, this is a regular practice done seasonally. Nevertheless, the hydrological role of these geomorphological features has to be considered, if more permanent measures are to be taken. Permanent removal of the sand ridges will negatively impact the water balance of the mangrove forest, allowing loss of much needed fresh water flows during the dry season. Fresh water inputs regulate mangroves growth and mortality. Changes in the hydrological regime would alter the structure and function of this ecosystem, which provides various social, economic and ecological services.

Sand ridges regulate river stage increases during rainfall events and also during dry periods. Therefore, they also control river–aquifer interactions. Groundwater recharge from river water is crucial during dry periods, especially considering the dependence of the local community on shallow groundwater resources. Tidal sand ridges are also a natural barrier against sea level rise and associated flooding. In view of climate change projections for the region for the mid-21st century, this is also an important consideration for the sustainable water resources management of the catchment and preservation of key ecosystems.

7.1.4. Hydrology and ecosystems

The catchment response to the hydro-climatic and geomorphologic controls supports the mangrove ecosystem freshwater needs. Surface water accumulation caused by the sand ridges enables the ecosystem to function and survive through dry periods. These geomorphologic features in the study area control freshwater and seawater mixing and in/outflows. During low flow conditions, the ridges prevent river discharge into the ocean, causing flooding of the mangrove forest and groundwater and surface water accumulation. This is observed as increased groundwater levels. Water accumulation helps to maintain a positive water balance for the mangrove forest during dry periods, despite high evaporation and groundwater discharge fluxes. Freshwater availability regulates growth, mortality, and phenological events. Changes in hydrological conditions would result in drastic alteration of structural and functional characteristics of the forest.

The mangrove forest in turn, provides a natural defense against flooding, habitat for numerous species, socio-economic benefits as a touristic attraction, and acts as a nutrient sink. The mangrove ecosystems on the South Pacific Coast are part of the biological corridor of the Pacific of Nicaragua, providing biological interconnectivity with other ecosystems. Mangroves are important ecosystems within these catchments and are very vulnerable to the changes in the hydrologic regime. Therefore, sustainable water resources management should carefully consider their protection and sustainable use.

7.1.5. Implications for other catchments

The water resources management and ecological implications of this study are also applicable to other catchments in this region. The hydro-climatic and structural characteristics of the Ostional catchment are similar to other catchments on the South Pacific Coast of Nicaragua. Regional geology, land use, and topography are similar. Mangrove forests are also common in this region and the presence of sand ridges is characteristic on the Pacific Coast of Central America. Thus, protection of mangrove forests in this region requires safeguarding the sand ridges. Furthermore, the high spatial and temporal variability of precipitation in this small catchment highlights the need to carefully monitor hydro-climatic parameters in other catchments of the region.

Additionally, the combination of non-invasive, integrative methods used in this work can also be applied to other catchments in Central America, where lack of monitoring networks and accessibility are challenging as well. This methodological approach is efficient and practical for remote and geological complex areas.

7.2.   Recommendations for future research

The South Pacific Coast of Nicaragua is very suitable for tourism and real estate development, which implies larger stress on water resources in this region in the near future. Additionally, the Interoceanic Grand Canal project will intervene the Southwestern Pacific Coast; and its sustainable development requires a sound hydrological understanding of the catchments involved. This understanding can only be achieved by developing means of comprehensive hydrometeorological data collection and assigning human and financial resources to hydrological research. Therefore, it is necessary to implement hydrological process studies in key regions which help obtain knowledge transferrable to other catchments and be able to establish the links between climate, catchment form and catchment response. The recommendations provided here are also applicable to other catchments in this region of Nicaragua.

Future research in this catchment should include: i) long term hydro-climatic monitoring, ii) water quality monitoring, iii) river-aquifer interactions and, iv) moisture recycling in forested areas, v) hillslope processes and, vi) mangrove conservation.

Long term hydro-climatic monitoring

Long term records of precipitation, evaporation, solar radiation, wind speed, and stream discharge are necessary to analyze long term variations in the water balance of the catchment.

This is crucial to guarantee water supply for social and economic development, and also to prevent ecosystem degradation.

It is essential to construct a gauging station to develop a rating curve, which considers continuous morphological changes in the stream channel. The design of a stream discharge monitoring station should consider bankfull as well as extreme events which may cause material damages and data loss. The piezometers installed in the lower catchment area should be used to continue monitoring groundwater level fluctuations and water chemistry at different depths.

The groundwater level monitoring network should be extended where possible. The piezometer network near the catchment outlet should be extended. However, remoteness and accessibility issues increase costs of monitoring in the upper catchment area, where springs are located. Thus, a different strategy should be implemented here. Areas where groundwater intersects the ground surface (springs) should be used as monitoring points. Chemical and isotopic tracers can provide valuable information of water sources and residence times. Seasonal changes in water chemistry and isotopic content are also important sources of information to understand runoff generation processes.

Water quality

Water quality of the river and groundwater should be assessed on a regular basis. The lack of waste water treatment facilities and aquifer recharge from river water pose threats to groundwater quality and human health. The effects of the nutrient input from waste water from the Ostional town into the mangrove forest needs further investigation. Although the forest is possible a nutrient sink, and thus removes nutrients from waste waters, it could be also affected by excessive nutrient inputs. This asks for further detailed studies.

Innovative tracers

Bacterial DNA as a natural tracer is a promising tool for rainfall-runoff process study. However, the experimental applications at field scale has to be increased to produce reliable results. Such experiments should consider different structural catchment characteristics and hydro-climatic conditions in order to compare them and test its capabilities under different conditions. Characteristics such as soil type and land use should be considered regarding suspended sediments and possibly qPCR inhibitory substances. The applied on site DNA harvesting method is still in need of improvements, especially during the sample pre-filtration steps.

River-aquifer interactions

The effect of bankfull events and overflow events on aquifer recharge and groundwater quality should be further investigated. Longer time series of river stage–groundwater level data combined with hydrochemical and isotopic information will improve analysis of these interactions, especially during highly variable hydro-climatic conditions (i.e. extreme precipitation events, flooding). Ecohydrological implications of these interactions should also be investigated. A model will help improve process analysis by testing hypothesis.

Moisture recycling

Moisture recycling in the forested sub-catchments needs further research. Given that the South Pacific Coast of Nicaragua is an isthmus between the Pacific Ocean and Lake Nicaragua, it is necessary to investigate the influence of lake water evaporation on the precipitation regime of the region. Stable water isotopic monitoring of precipitation is necessary to determine if re-evaporation from interception or from soil moisture is occurring.

Water isotopic composition from precipitation, water from Lake Nicaragua, soil water and intercepted water should be monitored to determine d-excess. This information will improve our understanding of the role of the forested areas in the hydrological response of the catchment. It will also provide more information for land use management of the area and protection of the runoff generation and groundwater recharge areas.

Hillslope processes

Subsurface stormflow is probably delivered to the stream through the soil matrix and macropores/pipes processes. However, detailed process examination needs more intense small scale investigations including soil moisture patterns. It is recommended that the Nancite sub-catchment (S11) be selected to continue detailed process studies. This sub-catchment is representative of the upper area and it is more accessible than the other forested sub-catchments. Additionally, the independent variations in stable water isotopes reflected in sub-catchment scale hydrograph separations may indicate various sources of atmospheric moisture generating precipitation. Stable water isotope enrichment by throughfall needs also to be further investigated.

Mangrove conservation

The mangrove forest ecosystem should be ecologically characterized to provide in-depth information on mangrove species, zonation patterns, effects of soil water salinity changes on the mangrove health and its relation to freshwater fluxes. The environmental flows needed to maintain this important ecosystem should be determine, as this information will be pertinent for other mangrove forest in the South Pacific of the country.

References

Abbott, M. D., Lini, A. and Bierman, P. R. 2000. δ18O, δD and 3H measurements constrain groundwater recharge patterns in an upland fractured bedrock aquifer, Vermont, USA. *Journal of Hydrology,* 228(1–2), 101-112.

Aleström, P., 1995. *Novel method for chemical labeling of objects.* application Int Patent Applic. no. PCT/IB95/01 144.

Allen, R. G., Pereira, L. S., Raes, D. and Smith, M. 1998. Crop evapotranspiration. Guidelines for computing crop water requirements-FAO Irrigation and drainage paper 56. *FAO, Rome,* 300, 6541.

Alley, W. M., Healy, R. W., LaBaugh, J. W. and Reilly, T. E. 2002. Flow and Storage in Groundwater Systems. *Science,* 296(5575), 1985-1990.

Alley, W. M., La Baugh, J. W. and Reilly, T. E., 2006. Groundwater as an Element in the Hydrological Cycle. *In:* Anderson, M. G. and McDonnell, J. J. eds. *Encyclopedia of Hydrological Sciences.* John Wiley & Sons, Ltd, 1-12.

Alongi, D. M. 2002. Present state and future of the world's mangrove forests. *Environmental Conservation,* 29(3), 331-349.

Amador, J. A., Alfaro, E. J., Lizano, O. G. and Magaña, V. O. 2006. Atmospheric forcing of the eastern tropical Pacific: A review. *Progress in Oceanography,* 69(2–4), 101-142.

Anderson, M. P. 2005. Heat as a Ground Water Tracer. *Ground Water,* 43(6), 951-968.

Andréassian, V. 2004. Waters and forests: from historical controversy to scientific debate. *Journal of Hydrology,* 291(1–2), 1-27.

Andréassian, V., Perrin, C., Parent, E. and Bárdossy, A. 2010. The Court of Miracles of Hydrology: can failure stories contribute to hydrological science? *Hydrological Sciences Journal,* 55(6), 849-856.

APHA, 1998. Standard methods for the examination of water and wastewater. *Standard methods for the examination of water and wastewater.* 20th ed. Washington, DC 156-158.

Appelo, C. and Postma, D., 1993. *Geochemistry, groundwater and pollution.* Rotterdam: AA Balkema.

Aquilanti, L., Clementi, F., Landolfo, S., Nanni, T., Palpacelli, S. and Tazioli, A. 2013. A DNA tracer used in column tests for hydrogeology applications. *Environmental Earth Sciences,* 1-12.

Araguas, L., 1992. *Estudio de Hidrología Isotópica de los Acuíferos de Managua (Study of the isotopic hydrology of the Managua aquifers).* OIEA/INETER.

ASTM, 1970. *Special procedures for testing soil and rock for engineering purposes.* 5th ed. Maryland: ASTM International.

Aylward, B., 2004. Land use, hydrological function and economic valuation. *In:* Bonell, M. and Bruijnzeel, L. A. eds. *Forests, Water and People in the Humid Tropics: Past, Present and Future Hydrological Research for Integrated Land and Water Management.* UK: Cambridge University Press, 99-120.

Banks, E., Simmons, C., Love, A., Cranswick, R., Werner, A., Bestland, E., Wood, M. and Wilson, T. 2009. Fractured bedrock and saprolite hydrogeologic controls on groundwater/surface-water interaction: a conceptual model (Australia). *Hydrogeology Journal,* 1(21).

Barlow, P. M., DeSimone, L. A. and Moench, A. F. 2000. Aquifer response to stream-stage and recharge variations. II. Convolution method and applications. *Journal of Hydrology,* 230(3–4), 211-229.

Bartsch, S., Frei, S., Ruidisch, M., Shope, C. L., Peiffer, S., Kim, B. and Fleckenstein, J. H. 2014. River-aquifer exchange fluxes under monsoonal climate conditions. *Journal of Hydrology,* 509(0), 601-614.

Beck, H. E., Bruijnzeel, L. A., van Dijk, A. I. J. M., McVicar, T. R., Scatena, F. N. and Schellekens, J. 2013. The impact of forest regeneration on streamflow in 12 mesoscale humid tropical catchments. *Hydrol. Earth Syst. Sci.,* 17(7), 2613-2635.

Benegas, L., Jiménez, F., Faustino, J. and Gentes, I. 2008. Experiencias y desafíos para la cogestión de cuencas hidrográficas en América Latina (Experiences and challenges for watershed management in Latin America). *Conclusiones del seminario internacional. Recursos Naturales y Ambiente,* 55, 129-133.

Berkowitz, B. 2002. Characterizing flow and transport in fractured geological media: A review. *Advances in Water Resources,* 25(8–12), 861-884.

Bethune, D. N., Farvolden, R. N., Ryan, M. C. and Guzman, A. L. 1996. Industrial Contamination of a Municipal Water-Supply Lake by Induced Reversal of Ground-Water Flow, Managua, Nicaragua. *Ground Water,* 34(4), 699-708.

Blöschl, G. 2006. Hydrologic synthesis: Across processes, places, and scales. *Water Resources Research,* 42(3), W03S02.

Blume, T., Zehe, E. and Bronstert, A. 2007. Rainfall—runoff response, event-based runoff coefficients and hydrograph separation. *Hydrological Sciences Journal,* 52(5), 843-862.

Blume, T., Zehe, E. and Bronstert, A. 2008a. Investigation of runoff generation in a pristine, poorly gauged catchment in the Chilean Andes II: Qualitative and quantitative use of tracers at three spatial scales. *Hydrological Processes,* 22(18), 3676-3688.

Blume, T., Zehe, E., Reusser, D. E., Iroumé, A. and Bronstert, A. 2008b. Investigation of runoff generation in a pristine, poorly gauged catchment in the Chilean Andes I: A multi-method experimental study. *Hydrological Processes,* 22(18), 3661-3675.

Bogaard, T., Malet, J. P. and Schmittbuhl, J. 2012. Hydrological behaviour of unstable clay-shales slopes: the value of cross-disciplinary and multitechnological research at different scales. *Hydrological Processes,* 26(14), 2067-2070.

Bohté, R., Mul, M., Bogaard, T., Savenije, H., Uhlenbrook, S. and Kessler, T. 2010. Hydrograph separation and scale dependency of natural tracers in a semi-arid catchment. *Hydrology and Earth System Sciences Discussions,* 7(1), 1343-1372.

Bonell, M. 1993. Progress in the understanding of runoff generation dynamics in forests. *Journal of Hydrology,* 150(2–4), 217-275.

Bonell, M. 1998. Selected challenges in runoff generation research in forests from hillslope to headwater drainage basin scale 1. *JAWRA Journal of the American Water Resources Association,* 34(4), 765-785.

Bonell, M. and Bruijnzeel, L. A., 2004. *Forests, water and people in the humid tropics: past, present and future hydrological research for integrated land and water management.* UK: Cambridge University Press.

Boom, R., Sol, C., Beld, M., Weel, J., Goudsmit, J. and Wertheim-van Dillen, P. 1999. Improved silica-guanidiniumthiocyanate DNA isolation procedure based on selective binding of bovine alpha-casein to silica particles. *Journal of clinical microbiology,* 37(3), 615-619.

Boom, R., Sol, C., Salimans, M., Jansen, C., Wertheim-van Dillen, P. and Van der Noordaa, J. 1990. Rapid and simple method for purification of nucleic acids. *Journal of clinical microbiology,* 28(3), 495-503.

Borgia, A. and van Wyk de Vries, B. 2003. The volcano-tectonic evolution of Concepción, Nicaragua. *Bulletin of Volcanology,* 65(4), 248-266.

Bouwer, H. and Rice, R. 1976. A slug test for determining hydraulic conductivity of unconfined aquifers with completely or partially penetrating wells. *Water Resources Research,* 12(3), 423-428.

Bowen, G. J., Wassenaar, L. I. and Hobson, K. A. 2005. Global application of stable hydrogen and oxygen isotopes to wildlife forensics. *Oecologia,* 143(3), 337-348.

Briceño, H., Miller, G. and Davis, S. E. 2013. Relating freshwater flow with estuarine water quality in the Southern Everglades mangrove ecotone. *Wetlands,* 1-11.

Brooks, J. R., Barnard, H. R., Coulombe, R. and McDonnell, J. J. 2010. Ecohydrologic separation of water between trees and streams in a Mediterranean climate. *Nature Geosci,* 3(2), 100-104.

Bruijnzeel, L. 2001. Hydrology of tropical montane cloud forests: a reassessment. *Land use and water resources research,* 1(1), 1.

Bruijnzeel, L., 2004a. Tropical montane cloud forest: a unique hydrological case. *In:* Bonell, M. and Bruijnzeel, L. A. eds. *Forests, Water and People in the Humid Tropics: Past, Present and Future Hydrological Research for Integrated Land and Water Management.* UK: Cambridge University Press, 462-483.

Bruijnzeel, L. A. 2004b. Hydrological functions of tropical forests: not seeing the soil for the trees? *Agriculture, Ecosystems & Environment,* 104(1), 185-228.

Bruijnzeel, L. A., Mulligan, M. and Scatena, F. N. 2011. Hydrometeorology of tropical montane cloud forests: emerging patterns. *Hydrological Processes,* 25(3), 465-498.

Brunke, M. and Gonser, T. 1997. The ecological significance of exchange processes between rivers and groundwater. *Freshwater Biology,* 37(1), 1-33.

Buchanan, T. and Somers, W., 1969. Discharge measurements at gaging stations. *Techniques of water-resources investigations of the USGS.* Washington, DC, 1-4.

Burns, D. 2002. Stormflow-hydrograph separation based on isotopes: the thrill is gone—what's next? *Hydrological Processes,* 16(7), 1515-1517.

Burt, T., Butcher, D., Coles, N. and Thomas, A. 1983. The natural history of Slapton Ley Nature Reserve XV: hydrological processes in the Slapton Wood catchment. *Field Studies,* 5(5), 731-732.

Caballero, L. A., 2012. *Hydrology, hydrochemistry and implications for water supply of a cloud forest in Central América.* (PhD). Cornell University.

Caballero, L. A., Easton Zachary, M., Richards Brian, K. and Steenhuis Tammo, S. 2013. Evaluating the bio-hydrological impact of a cloud forest in Central America using a semi-distributed water balance model. *Journal of Hydrology and Hydromechanics,* 61(1), 9.

Cahoon, D. R., Hensel, P., Rybczyk, J., McKee, K. L., Proffitt, C. E. and Perez, B. C. 2003. Mass tree mortality leads to mangrove peat collapse at Bay Islands, Honduras after Hurricane Mitch. *Journal of ecology,* 91(6), 1093-1105.

Calderon, H., 2010. *Dictamen científico técnico de la actual Política Nacional de los Recursos Hídricos (Scientific and technical analysis of the National Water Resources Policy).* Managua: National Water Authority.

Calderon, H. and Bentley, L. R. 2007. A regional-scale groundwater flow model for the Leon-Chinandega aquifer, Nicaragua. *Hydrogeology Journal,* 15(8), 1457-1472.

Calderon, H. and Flores, Y. 2011. Evaluación de la dinámica de la laguna de Apoyo mediante trazadores isotópicos y geoquímicos (Assessment of the dynamics of the Apoyo lake using isotopic and geochemical tracers). *Universidad y Ciencia,* 5(8), 22-26.

Calderon, H., Flores, Y., Corriols, M., Sequeira, L. and Uhlenbrook, S. in review. Integrating geophysical, tracer and hydrochemical data to conceptualize groundwater flow systems in a tropical coastal catchment. *Environmental Earth Sciences.*

Calderon, H. and Uhlenbrook, S. 2014a. Characterising the climatic water balance dynamics and different runoff components in a poorly gauged tropical forested catchment, Nicaragua. *Hydrological Sciences Journal.*

Calderon, H. and Uhlenbrook, S. 2014b. Investigation of seasonal river–aquifer interactions in a tropical coastal area controlled by tidal sand ridges. *Hydrology Earth System Science Discussions,* 11(8), 9759-9790.

Calderon, H. and Uhlenbrook, S. submitted. Lessons learned from catchment scale tracer test experiments using qPCR and natural DNA from total bacteria during rainfall-runoff events in a tropical environment. *Hydrological Processes.*

Calderon, H., Weeda, R. and Uhlenbrook, S. 2014. Hydrological and geomorphological controls on the water balance components of a mangrove forest during the dry season in the Pacific Coast of Nicaragua. *Wetlands,* 34(4), 685-697.

Calero, S., Fomsgaard, I., Lacayo, M. L., Martinez, V. and Rugama, R. 1993. Toxaphene and Other Organochlorine Pesticides in Fish and Sediment from Lake Xolotlán, Nicaragua. *International Journal of Environmental Analytical Chemistry,* 53(4), 297-305.

Calvo, J. C. 1986. An evaluation of Thornthwaite's water balance technique in predicting stream runoff in Costa Rica. *Hydrological Sciences Journal,* 31(1), 51-60.

Capell, R., Tetzlaff, D., Malcolm, I. A., Hartley, A. J. and Soulsby, C. 2011. Using hydrochemical tracers to conceptualise hydrological function in a larger scale catchment draining contrasting geologic provinces. *Journal of Hydrology,* 408(1–2), 164-177.

Carvalho, F., Montenegro-Guillén, S., Villeneuve, J., Cattini, C., Bartocci, J., Lacayo, M. and Cruz, A. 1999. Chlorinated hydrocarbons in coastal lagoons of the Pacific Coast of Nicaragua. *Archives of Environmental Contamination and Toxicology,* 36(2), 132-139.

Carvalho, F., Montenegro-Guillén, S., Villeneuve, J., Cattini, C., Tolosa, I., Bartocci, J., Lacayo-Romero, M. and Cruz-Granja, A. 2003. Toxaphene residues from cotton fields in soils and in the coastal environment of Nicaragua. *Chemosphere,* 53(6), 627-636.

Castañeda-Moya, E., Rivera-Monroy, V. H. and Twilley, R. R. 2006. Mangrove zonation in the dry life zone of the Gulf of Fonseca, Honduras. *Estuaries and Coasts,* 29(5), 751-764.

Castilho, J. A. A., Fenzl, N., Montenergo Guillén, S. and Nascimento, F. S. 2000. Organochlorine and organophosphorus pesticide residues in the Atoya river basin, Chinandega, Nicaragua. *Environmental Pollution,* 110(3), 523-533.

Castillo Hernández, E., Calderón Palma, H., Delgado Quezada, V., Flores Meza, Y. and Salvatierra Suárez, T. 2006. Situación de los recursos hídricos en Nicaragua. *Boletín Geológico y Minero,* 117(1), 127-146.

Cavelier, J., Jaramillo, M. a., Solis, D. and de León, D. 1997. Water balance and nutrient inputs in bulk precipitation in tropical montane cloud forest in Panama. *Journal of Hydrology,* 193(1–4), 83-96.

CBM, 2002. *El Corredor Biológico Mesoamericano: Una plataforma para el desarrollo sostenible regional (The Mesoamerican Biological Corridor: a platform for regional sustainable development).* Proyecto Corredor biológico mesoamericano.

Cey, E. E., Rudolph, D. L., Parkin, G. W. and Aravena, R. 1998. Quantifying groundwater discharge to a small perennial stream in southern Ontario, Canada. *Journal of Hydrology,* 210(1–4), 21-37.

Chaves, J., Neill, C., Germer, S., Neto, S. G., Krusche, A. and Elsenbeer, H. 2008. Land management impacts on runoff sources in small Amazon watersheds. *Hydrological Processes,* 22(12), 1766-1775.

Chen, X. and Chen, X. 2003. Stream water infiltration, bank storage, and storage zone changes due to stream-stage fluctuations. *Journal of Hydrology,* 280(1–4), 246-264.

Choza, A. 2002. Elementos básicos para la protección de las aguas subterráneas aplicados en el acuífero de Managua, Nicaragua (Basic elements for the protection of groundwater applied to the Managua aquifer, Nicaragua). *Revista Geológica de América Central,* 1(27), 61-74.

Chung, J. W., Foppen, J. W. and Lens, P. N. L. 2013. Development of Low Cost Two-Step Reverse Transcription-Quantitative Polymerase Chain Reaction Assays for Rotavirus Detection in Foul Surface Water Drains. *Food and Environmental Virology,* 5(2), 126-133.

CIRA, 2007. *Evaluación del estado actual (calidad y cantidad) y estimación de la dinámica de recarga de los mantos freáticos en la franja costera del Municipios de Tola (Assement of the quality, quantity and groundwater recharge dynamics in the coastal area of Tola).* Managua: Centro para la Investigación en Recursos Acuáticos de Nicaragua.

CIRA, 2008. *Disponibilidad actual y futura de los recursos hídricos en la franja costera del municipio de San Juan del Sur (Actual and future water resources availability in the coastal area of the municipality of San Juan del Sur).* Managua: Centro para la Investigación en Recursos Acuáticos de Nicaragua.

Clark, I. and Fritz, P., 1997. Environmental isotopes in hydrology. Boca Raton, Florida: Lewis Publisher.

Clesceri, L. S., Greenberg, A. E. and Eaton, A. D., 1998. *Standard Methods for the Examination of Water and Wastewater, 20th Edition.* Washington DC: APHA American Public Health Association.

Cloutier, C.-A., Buffin-Bélanger, T. and Larocque, M. 2014. Controls of groundwater floodwave propagation in a gravelly floodplain. *Journal of Hydrology,* 511(0), 423-431.

Cohen, S., Neilsen, D., Smith, S., Neale, T., Taylor, B., Barton, M., Merritt, W., Alila, Y., Shepherd, P. and Mcneill, R. 2006. Learning with local help: Expanding the dialogue on climate change and water management in the Okanagan Region, British Columbia, Canada. *Climatic change,* 75(3), 331-358.

Colleuille, H., Kitterod, N. and 2000, i. S. e. a. 1998. Forurensning av drikkevannsbronn pä Sundreoya IAl kommune. *Resultat av sporstoffforsok, Norwegian Water Resources and Energy Directorate Report,* 18.

Conant, B. 2004. Delineating and Quantifying Ground Water Discharge Zones Using Streambed Temperatures. *Ground Water,* 42(2), 243-257.

Corrales, D. and Delgado, V., 2009. Estudio del Acuífero Aluvial del Valle de Estelí, Nicaragua (Study of the alluvial aquifer of the Esteli valley, Nicaragua). *Estudios de Hidrología Isotópica en América Latina (Isotopic hydrology studies in Latin America).* Vienna, Austria: IAEA, 232.

Corredor, J. E. and Morell, J. M. 1994. Nitrate depuration of secondary sewage effluents in mangrove sediments. *Estuaries and Coasts,* 17(1), 295-300.

Corriols, M. and Dahlin, T. 2008. Geophysical characterization of the León-Chinandega aquifer, Nicaragua. *Hydrogeology Journal,* 16(2), 349-362.

Corriols, M., Nielsen, M. R., Dahlin, T. and Christensen, N. 2009. Aquifer investigations in the León-Chinandega plains, Nicaragua, using electromagnetic and electrical methods. *Near Surface Geophysics,* 7(5-6), 413-425.

Costa, M., 2004. Large-scale hydrological impacts of tropical forest conversion. *In:* Bonell, M. and Bruijnzeel, L. A. eds. *Forests, Water and People in the Humid Tropics: Past, Present and Future Hydrological Research for Integrated Land and Water Management.* UK: Cambridge University Press, 590-595.

Cruz, O., 1997. *Modelaje del Acuífero de Managua y su Rendimiento Sostenible (Modeling of the Managua aquifer and its safe yield).* Master thesis. Universidad de Costa Rica.

Dahlin, T. 1996. 2D resistivity surveying for environmental and engineering applications. *First Break,* 14(7), 275-283.

Dahlin, T. and Linderman, J. E. 2007. Software Developed in Cooperation between ABEM. *Lund University and Terraohm.*

Dansgaard, W. 1964. Stable isotopes in precipitation. *Tellus,* 16(4), 436-468.

de Araújo, J. C. and González Piedra, J. I. 2009. Comparative hydrology: analysis of a semiarid and a humid tropical watershed. *Hydrological Processes,* 23(8), 1169-1178.

de Montety, V., Radakovitch, O., Vallet-Coulomb, C., Blavoux, B., Hermitte, D. and Valles, V. 2008. Origin of groundwater salinity and hydrogeochemical processes in a confined coastal aquifer: Case of the Rhône delta (Southern France). *Applied Geochemistry,* 23(8), 2337-2349.

Delgado Quezada, V., 2003. *Groundwater flow system and water quality in a coastal plain aquifer in northwestern Nicaragua.* Master thesis. University of Calgary.

Didszun, J. and Uhlenbrook, S. 2008. Scaling of dominant runoff generation processes: Nested catchments approach using multiple tracers. *Water Resources Research,* 44(2), W02410.

Diekkrüger, B. and Hiepe, C., 2012. The role of modeling for integrated water resource management. *In:* Bormann, H. and Althoff, I. eds. *Watershed Management and Rural Sanitation* Siegen: Universität Siegen, 63-78.

Dingman, S. L., 2002. *Physical Hydrology.* 2nd ed. New Jersey: Prentice Hall.

Dionisi, H. M., Harms, G., Layton, A. C., Gregory, I. R., Parker, J., Hawkins, S. A., Robinson, K. G. and Sayler, G. S. 2003. Power analysis for real-time PCR quantification of genes in activated sludge and analysis of the variability introduced by DNA extraction. *Applied and Environmental Microbiology,* 69(11), 6597-6604.

Durán-Quesada, A. M., Reboita, M. and Gimeno, L. 2012. Precipitation in tropical America and the associated sources of moisture: a short review. *Hydrological Sciences Journal,* 57(4), 612-624.

Ellison, A. 2004. Wetlands of Central America. *Wetlands Ecology and Management,* 12(1), 3-55.

Elming, S. Å., Layer, P. and Ubieta, K. 2001. A palaeomagnetic study and age determinations of Tertiary rocks in Nicaragua, Central America. *Geophysical Journal International,* 147(2), 294-309.

Elming, S. Å. and Rasmussen, T. 1997. Results of magnetotelluric and gravimetric measurements in western Nicaragua, Central America. *Geophysical Journal International,* 128(3), 647-658.

Elsenbeer, H. 2001. Hydrologic flowpaths in tropical rainforest soilscapes—a review. *Hydrological Processes,* 15(10), 1751-1759.

Elsenbeer, H., Lack, A. and Cassel, K. 1995a. Chemical fingerprints of hydrological compartments and flow paths at La Cuenca, Western Amazonia. *Water Resources Research,* 31(12), 3051-3058.

Elsenbeer, H., Lorieri, D. and Bonell, M. 1995b. Mixing Model Approaches to Estimate Storm Flow Sources in an Overland Flow-Dominated Tropical Rain Forest Catchment. *Water Resources Research,* 31(9), 2267-2278.

Elsenbeer, H. and Vertessy, R. A. 2000. Stormflow generation and flowpath characteristics in an Amazonian rainforest catchment. *Hydrological Processes,* 14(14), 2367-2381.

Falkenmark, M. 2004. Towards integrated catchment management: opening the paradigm locks between hydrology, ecology and policy-making. *International Journal of Water Resources Development,* 20(3), 275-281.

FAO, 2010. *Global Forest Resources Assessment 2010 Main Report.* Rome: FAO.

Faustino, J. and García, S. 2001. Manejo de cuencas hidrográficas (Watershed management). *Conceptos, gestión, planificación, implementación y monitoreo. San Salvador, El Salvador.*

Faustino, J., Jiménez, F. and Kammerbaeur, H. 2007. La cogestión de cuencas hidrográficas en América Central: Planteamiento conceptual y experiencias de implementación (Co-management of watersheds in Central America: Coceptual set up and implementation). *Turrialba, CR, CATIE/Asdi.*

Feller, I. C. and Sitnik, M. 1996. Mangrove ecology workshop manual. *Smithsonian Institution, Washington, DC*, 135.

Fenzl, N., 1989. *Nicaragua: Geografía, clima, geología e hidrogeología (Nicaragua: Geography, climate, geology and hydrogeology).* UFPA/INETER/INAN.

Ferris, J. G., Knowles, D., Brown, R. and Stallman, R. W., 1962. *Theory of aquifer tests.* US Geological Survey.

Fetter, C., 2001. *Applied hydrogeology.* 4th ed. New Jersey: Prentice Hall Upper Saddle River.

Fetter, C. W., 1999. *Contaminant hydrogeology.* Prentice Hall Upper Saddle River, New Jersey.

Foppen, J. W., Orup, C., Adell, R., Poulalion, V. and Uhlenbrook, S. 2011. Using multiple artificial DNA tracers in hydrology. *Hydrological Processes,* 25(19), 3101-3106.

Foppen, J. W., Seopa, J., Bakobie, N. and Bogaard, T. 2013. Development of a methodology for the application of synthetic DNA in stream tracer injection experiments. *Water Resources Research*, n/a-n/a.

Foster, S. and MacDonald, A. 2014. The 'water security' dialogue: why it needs to be better informed about groundwater. *Hydrogeology Journal*, 1-4.

Francese, R., Mazzarini, F., Bistacchi, A., Morelli, G., Pasquarè, G., Praticelli, N., Robain, H., Wardell, N. and Zaja, A. 2009. A structural and geophysical approach to the study of fractured aquifers in the Scansano-Magliano in Toscana Ridge, southern Tuscany, Italy. *Hydrogeology Journal,* 17(5), 1233-1246.

Fürst, E., Barton, D. N. and Jimenez, G., 2000. Partial economic valuation of mangroves in Nicaragua *In:* McCracken, J. R. and Abaza, H. eds. *Environmental valuation: a worldwide compendium of case studies.* UK: Earthscan, 233.

Gaál, L., Szolgay, J., Kohnová, S., Parajka, J., Merz, R., Viglione, A. and Blöschl, G. 2012. Flood timescales: Understanding the interplay of climate and catchment processes through comparative hydrology. *Water Resources Research,* 48(4), W04511.

Genereux, D. 2004. Comparison of naturally-occurring chloride and oxygen-18 as tracers of interbasin groundwater transfer in lowland rainforest, Costa Rica. *Journal of Hydrology,* 295(1–4), 17-27.

Genereux, D. P. and Hooper, R. P. 1998. Oxygen and hydrogen isotopes in rainfall-runoff studies. *Isotope tracers in catchment hydrology*, 319-346.

George, R. K., Waylen, P. and Laporte, S. 1998. Interannual variability of annual streamflow and the Southern Oscillation in Costa Rica. *Hydrological Sciences Journal,* 43(3), 409-424.

Germer, S., Neill, C., Krusche, A. V. and Elsenbeer, H. 2010. Influence of land-use change on near-surface hydrological processes: Undisturbed forest to pasture. *Journal of Hydrology,* 380(3–4), 473-480.

Germer, S., Neill, C., Vetter, T., Chaves, J., Krusche, A. V. and Elsenbeer, H. 2009. Implications of long-term land-use change for the hydrology and solute budgets of small catchments in Amazonia. *Journal of Hydrology,* 364(3–4), 349-363.

Ghiglieri, G., Carletti, A. and Pittalis, D. 2012. Analysis of salinization processes in the coastal carbonate aquifer of Porto Torres (NW Sardinia, Italy). *Journal of Hydrology,* 432–433(0), 43-51.

Giertz, S., Junge, B. and Diekkrüger, B. 2005. Assessing the effects of land use change on soil physical properties and hydrological processes in the sub-humid tropical environment of West Africa. *Physics and Chemistry of the Earth, Parts A/B/C,* 30(8–10), 485-496.

Gilman, E. L., Ellison, J., Duke, N. C. and Field, C. 2008. Threats to mangroves from climate change and adaptation options: a review. *Aquatic Botany,* 89(2), 237-250.

Giménez, E. and Morell, I. 1997. Hydrogeochemical analysis of salinization processes in the coastal aquifer of Oropesa (Castellón, Spain). *Environmental Geology,* 29(1-2), 118-131.

Giorgi, F. 2006. Climate change hot-spots. *Geophysical Research Letters,* 33(8), L08707.

Glynn, P. D. and Plummer, L. N. 2005. Geochemistry and the understanding of ground-water systems. *Hydrogeology Journal,* 13(1), 263-287.

Goller, R., Wilcke, W., Leng, M. J., Tobschall, H. J., Wagner, K., Valarezo, C. and Zech, W. 2005. Tracing water paths through small catchments under a tropical montane rain forest in south Ecuador by an oxygen isotope approach. *Journal of Hydrology,* 308(1–4), 67-80.

Gondwe, B. N., Hong, S.-H., Wdowinski, S. and Bauer-Gottwein, P. 2010. Hydrologic dynamics of the ground-water-dependent Sian Ka'an wetlands, Mexico, derived from InSAR and SAR data. *Wetlands,* 30(1), 1-13.

Gooch, G. D., 2010. *Science, Policy, and Stakeholders in Water Management: An Integrated Approach to River Basin Management.* Earthscan.

Groombridge, B., 1992. *Global biodiversity: status of the earth's living resources.* Chapman & Hall.

Gross, J., Flores, E. and Schwendenmann, L. 2013. Stand structure and aboveground biomass of a *Pelliciera rhizophorae* mangrove forest, Gulf of Monitjo Ramsar site, Pacific Coast, Panama. *Wetlands,* 1-11.

Guerrero, J.-L., Westerberg, I. K., Halldin, S., Xu, C.-Y. and Lundin, L.-C. 2012. Temporal variability in stage–discharge relationships. *Journal of Hydrology,* 446–447(0), 90-102.

GWP, 2011. *Situación de los recursos hídricos en Centroamérica (State of the water resources in Central America).* Honduras.

Häggström, M., Lindström, G., Cobos, C., Martinez, J. R., Merlos, L., Alonzo, R. D., Castillo, G., Sirias, C., Miranda, D. and Granados, J., 1990. *Application of the HBV model for flood forecasting in six Central American rivers.* Sweden: SMHI Norrköping.

Harmon, R. S., Berry Lyons, W., Long, D. T., Ogden, F. L., Mitasova, H., Gardner, C. B., Welch, K. A. and Witherow, R. A. 2009. Geochemistry of four tropical montane watersheds, Central Panama. *Applied Geochemistry,* 24(4), 624-640.

Harms, G., Layton, A. C., Dionisi, H. M., Gregory, I. R., Garrett, V. M., Hawkins, S. A., Robinson, K. G. and Sayler, G. S. 2003. Real-time PCR quantification of nitrifying bacteria in a municipal wastewater treatment plant. *Environmental science & technology,* 37(2), 343-351.

Hassan, A., Velasquez, E., Belmar, R., Coye, M., Drucker, E., Landrigan, P. J., Michaels, D. and Sidel, K. B. 1981. Mercury poisoning in Nicaragua. *International Journal of Health Services,* 11(2), 221-226.

Hastenrath, S. L. 1967. Rainfall distribution and regime in Central America. *Archiv für Meteorologie, Geophysik und Bioklimatologie, Serie B,* 15(3), 201-241.

Hata, A., Katayama, H., Kitajima, M., Visvanathan, C., Nol, C. and Furumai, H. 2011. Validation of Internal Controls for Extraction and Amplification of Nucleic Acids from Enteric Viruses in Water Samples. *Applied and Environmental Microbiology,* 77(13), 4336-4343.

Hata, A., Katayama, H., Kojima, K., Sano, S., Kasuga, I., Kitajima, M. and Furumai, H. 2014. Effects of rainfall events on the occurrence and detection efficiency of viruses in river water impacted by combined sewer overflows. *Science of The Total Environment,* 468–469(0), 757-763.

He, G., Engel, V., Leonard, L., Croft, A., Childers, D., Laas, M., Deng, Y. and Solo-Gabriele, H. 2010. Factors Controlling Surface Water Flow in a Low-gradient Subtropical Wetland. *Wetlands,* 30(2), 275-286.

He, J., Shen, J., Zhang, L., Zhu, Y., Zheng, Y., Xu, M. and Di, H. 2007. Quantitative analyses of the abundance and composition of ammonia-oxidizing bacteria and ammonia-oxidizing archaea of a Chinese upland red soil under long-term fertilization practices. *Environmental Microbiology,* 9(9), 2364-2374.

Heid, C. A., Stevens, J., Livak, K. J. and Williams, P. M. 1996. Real time quantitative PCR. *Genome research,* 6(10), 986-994.

Henry, S., Baudoin, E., López-Gutiérrez, J. C., Martin-Laurent, F., Brauman, A. and Philippot, L. 2004. Quantification of denitrifying bacteria in soils by nirK gene targeted real-time PCR. *Journal of Microbiological Methods,* 59(3), 327-335.

Herschy, R. W. and Fairbridge, R. W., 1998. *Encyclopedia of hydrology and water resources.* Springer.

Hidalgo, H. G., Amador, J. A., Alfaro, E. J. and Quesada, B. 2013. Hydrological climate change projections for Central America. *Journal of Hydrology,* 495(0), 94-112.

Higuchi, R., Fockler, C., Dollinger, G. and Watson, R. 1993. Kinetic PCR analysis: real-time monitoring of DNA amplification reactions. *Biotechnology,* 11, 1026-1030.

Hoeg, S., Uhlenbrook, S. and Leibundgut, C. 2000. Hydrograph separation in a mountainous catchment — combining hydrochemical and isotopic tracers. *Hydrological Processes,* 14(7), 1199-1216.

Holder, C. D. 2004. Rainfall interception and fog precipitation in a tropical montane cloud forest of Guatemala. *Forest Ecology and Management,* 190(2–3), 373-384.

Holdridge, L. R., 1967. *Life zone ecology*. Revised ed. Costa Rica: Tropical Science Center.

Hölscher, D., Köhler, L., Leuschner, C. and Kappelle, M. 2003. Nutrient fluxes in stemflow and throughfall in three successional stages of an upper montane rain forest in Costa Rica. *Journal of Tropical Ecology,* 19(05), 557-565.

Hölscher, D., Köhler, L., van Dijk, A. I. J. M. and Bruijnzeel, L. A. 2004. The importance of epiphytes to total rainfall interception by a tropical montane rain forest in Costa Rica. *Journal of Hydrology,* 292(1–4), 308-322.

Hooper, R. P., Christophersen, N. and Peters, N. E. 1990. Modelling streamwater chemistry as a mixture of soilwater end-members — An application to the Panola Mountain catchment, Georgia, U.S.A. *Journal of Hydrology,* 116(1–4), 321-343.

Hoshino, T., Noda, N., Tsuneda, S., Hirata, A. and Inamori, Y. 2001. Direct Detection by In Situ PCR of theamoA Gene in Biofilm Resulting from a Nitrogen Removal Process. *Applied and Environmental Microbiology,* 67(11), 5261-5266.

House, W. A. and Warwick, M. S. 1998. Hysteresis of the solute concentration/discharge relationship in rivers during storms. *Water Research,* 32(8), 2279-2290.

Hrachowitz, M., Bohte, R., Mul, M., Bogaard, T., Savenije, H. and Uhlenbrook, S. 2011. On the value of combined event runoff and tracer analysis to improve understanding of catchment functioning in a data-scarce semi-arid area. *Hydrology and Earth System Sciences,* 15(6), 2007.

Hrachowitz, M., Savenije, H. H. G., Blöschl, G., McDonnell, J. J., Sivapalan, M., Pomeroy, J. W., Arheimer, B., Blume, T., Clark, M. P., Ehret, U., Fenicia, F., Freer, J. E., Gelfan, A., Gupta, H. V., Hughes, D. A., Hut, R. W., Montanari, A., Pande, S., Tetzlaff, D., Troch, P. A., Uhlenbrook, S., Wagener, T., Winsemius, H. C., Woods, R. A., Zehe, E. and Cudennec, C. 2013. A decade of Predictions in Ungauged Basins (PUB)—a review. *Hydrological Sciences Journal,* 1-58.

Hugenschmidt, C., Ingwersen, J., Sangchan, W., Sukvanachaikul, Y., Duffner, A., Uhlenbrook, S. and Streck, T. 2014. A three-component hydrograph separation based on geochemical tracers in a tropical mountainous headwater catchment in northern Thailand. *Hydrolgy and Earth System Sciences,* 18(2), 525-537.

Hugenschmidt, C., Ingwersen, J., Sangchan, W., Sukvanachaikul, Y., Uhlenbrook, S. and Streck, T. 2010. Hydrochemical analysis of stream water in a tropical, mountainous headwater catchment in northern Thailand. *Hydrology and Earth System Sciences Discussions,* 7(2), 2187-2220.

INETER, 2005. *Mapificacion hidrogeológica e hidroquímica de la región central de Nicaragua (Hydrogeologic and hydrochemical mapping of the central region of Nicaragua).* Managua: INETER.

INETER, 2009. *Caracterizacion hidrogeológica e isotópica del lago de Nicaragua (Hydrogeologic and isotopic characterization of Lake Nicaragua).* Managua: INETER.

Jeanson, M., Anthony, E., Dolique, F. and Cremades, C. 2014. Mangrove Evolution in Mayotte Island, Indian Ocean: A 60-year Synopsis Based on Aerial Photographs. *Wetlands,* 1-10.

Jiang, J., Alderisio, K. A., Singh, A. and Xiao, L. 2005. Development of procedures for direct extraction of Cryptosporidium DNA from water concentrates and for relief of PCR inhibitors. *Applied and Environmental Microbiology,* 71(3), 1135-1141.

Jimenez, J. 1990. The structure and function of dry weather mangroves on the Pacific Coast of Central America, with emphasis on *Avicennia bicolor* forests. *Estuaries,* 13(2), 182-192.

Jimenez, J. 1992. Mangrove forests of the Pacific coast of Central America. *Coastal plant communities of Latin America,* 259-267.

Jimenez, J., Yañez-Arancibia, A. and Lara-Domínguez, A., 1999. Ambiente, distribución y características estructurales en los manglares del Pacífico de Centro América: contrastes climáticos (Environment, distribution and structural characteristics of mangrove forests in Central America: climatic contrasts). *In:* Yañez-Arancibia, A. and Lara-Domínguez, A. eds. *Ecosistemas de manglar en América Tropical.* 1st ed. Mexico: Instituto de Ecología, A.C. Centro SEP - CONACYT, 51-70.

Johansson, P. O., Scharp, C., Alveteg, T. and Choza, A. 1999. Framework for Ground Water Protection - the Managua Ground Water System as an Example. *Ground Water,* 37(2), 204-213.

Jung, M., Burt, T. P. and Bates, P. D. 2004. Toward a conceptual model of floodplain water table response. *Water Resources Research,* 40(12), W12409.

Kaimowitz, D., 2004. Useful myths and intractable truths: the politics of the link between forests and water in Central America. *In:* Bonell, M. and Bruijnzeel, L. A. eds. *Forests, Water and People in the Humid Tropics: Past, Present and Future Hydrological Research for Integrated Land and Water Management.* UK: Cambridge University Press, 86-98.

Kalbus, E., Reinstorf, F. and Schirmer, M. 2006. Measuring methods for groundwater – surface water interactions: a review. *Hydrol. Earth Syst. Sci.,* 10(6), 873-887.

Kaser, D. H., Binley, A., Heathwaite, A. L. and Krause, S. 2009. Spatio-temporal variations of hyporheic flow in a riffle-step-pool sequence. *Hydrological Processes,* 23(15), 2138-2149.

Keddy, P. A., 2010. *Wetland ecology: principles and conservation.* Second ed. UK: Cambridge University Press.

Kendall, C. and McDonnell, J. J., 1998. *Isotope tracers in catchment hydrology.* First ed. Netherlands: Elsevier.

Kirchner, J. W. 2003. A double paradox in catchment hydrology and geochemistry. *Hydrological Processes,* 17(4), 871-874.

Kjerfve, B., Lacerda, L., Rezende, C. E. and Ovalle, A. R. C., 1999. Hydrological and hydrogeochemical variations in mangrove ecosystems. *In:* Yañez-Arancibia, A. and Lara-Domínguez, A. eds. *Mangrove ecosystems in tropical America: structure, function and management.* 1st ed. Mexico: Instituto de Ecología, A.C. Centro SEP - CONACYT, 71-81.

Klaus, J. and McDonnell, J. J. 2013. Hydrograph separation using stable isotopes: Review and evaluation. *Journal of Hydrology,* 505(0), 47-64.

Koch, K., Wenninger, J., Uhlenbrook, S. and Bonell, M. 2009. Joint interpretation of hydrological and geophysical data: electrical resistivity tomography results from a process hydrological research site in the Black Forest Mountains, Germany. *Hydrological Processes,* 23(10), 1501-1513.

Kongo, V. M., Jewitt, G. P. W. and Lorentz, S. A. 2007. Establishing a catchment monitoring network through a participatory approach: a case study from the Potshini catchment in the Thukela River basin, South Africa. *IWMI Working Paper,* (120).

Krasny, J. and Hecht, G., 1998. *Estudios hidrogeológicos e hidroquímicos de la Región del Pacífico de Nicaragua (Hydrogeologic and hydrochemical studies of the Pacific Region of Nicaragua).* Managua: INETER.

Krause, S., Blume, T. and Cassidy, N. 2012. Investigating patterns and controls of groundwater up-welling in a lowland river by combining fibre-optic distributed temperature sensing with observations of vertical head gradients. *Hydrology and Earth System Sciences Discussions,* 9(1), 337-378.

Kuang, J., 1971. *Estudio geológico del Pacífico de Nicaragua (Geological study of the Pacific of Nicaragua).* Managua: Catastro e Inventario de Recursos Naturales.

Kumpulainen, R. A. 1995. Stratigraphy and Sedimentology in Western Nicaragua. *Revista de Geologia de América Central,* 18, 91-94.

Kundzewicz, Z. W., Mata, L. J., Arnell, N. W., DÖLl, P., Jimenez, B., Miller, K., Oki, T., ŞEn, Z. and Shiklomanov, I. 2008. The implications of projected climate change for freshwater resources and their management. *Hydrological Sciences Journal,* 53(1), 3-10.

Lacayo, M., Cruz, A., Calero, S., Lacayo, J. and Fomsgaard, I. 1992. Total arsenic in water, fish, and sediments from Lake Xolotlán, Managua, Nicaragua. *Bulletin of Environmental Contamination and Toxicology,* 49(3), 463-470.

Lachniet, M. S. and Patterson, W. P. 2002. Stable isotope values of Costa Rican surface waters. *Journal of Hydrology,* 260(1–4), 135-150.

Leibundgut, C., Maloszewski, P. and Külls, C., 2009. Environmental Tracers. *Tracers in Hydrology.* John Wiley & Sons, Ltd, 13-56.

Lele, S. 2009. Watershed services of tropical forests: from hydrology to economic valuation to integrated analysis. *Current Opinion in Environmental Sustainability,* 1(2), 148-155.

Lewandowski, J., Lischeid, G. and Nützmann, G. 2009. Drivers of water level fluctuations and hydrological exchange between groundwater and surface water at the lowland River Spree (Germany): field study and statistical analyses. *Hydrological Processes,* 23(15), 2117-2128.

Ley General de Aguas Nacionales. Ley No. 620 (General Law of National Waters. Law No. 620) 2007.

Li, J., Li, B., Zhou, Y., Xu, J. and Zhao, J. 2011. A rapid DNA extraction method for PCR amplification from wetland soils. *Letters in applied microbiology,* 52(6), 626-633.

Llort, M. and Montufar, J. 2002. El Plan Trifinio y la cuenca compartida del río Lempa de El Salvador, Guatemala y Honduras. *Secretaria Ejecutiva Trinacional, San Salvador, El Salvador. p,* 247-264.

Loke, M., Acworth, I. and Dahlin, T. 2003. A comparison of smooth and blocky inversion methods in 2D electrical imaging surveys. *Exploration Geophysics,* 34(3), 182-187.

Loke, M. and Barker, R. 2004. RES2Dinv software. *Geotomo Software Company.*

Lorenz, M. G. and Wackernagel, W. 1987. Adsorption of DNA to sand and variable degradation rates of adsorbed DNA. *Applied and Environmental Microbiology,* 53(12), 2948-2952.

Lovelock, C., Feller, I. C., McKee, K. and Ball, M. 2004. The effect of nutrient enrichment on growth, photosynthesis and hydraulic conductance of dwarf mangroves in Panama. *Functional Ecology,* 18(1), 25-33.

Lugo, A. E. and Snedaker, S. C. 1974. The ecology of mangroves. *Annual Review of Ecology and Systematics,* 5(ArticleType: research-article / Full publication date: 1974 / Copyright © 1974 Annual Reviews), 39-64.

Lyon, S. W., Desilets, S. L. E. and Troch, P. A. 2009. A tale of two isotopes: differences in hydrograph separation for a runoff event when using δD versus δ18O. *Hydrological Processes,* 23(14), 2095-2101.

Lyon, S. W., Nathanson, M., Spans, A., Grabs, T., Laudon, H., Temnerud, J., Bishop, K. H. and Seibert, J. 2012. Specific discharge variability in a boreal landscape. *Water Resources Research,* 48(8), W08506.

Macinnis-Ng, C. M. O., Flores, E. E., Müller, H. and Schwendenmann, L. 2014. Throughfall and stemflow vary seasonally in different land-use types in a lower montane tropical region of Panama. *Hydrological Processes,* 28(4), 2174-2184.

MacNeil, R. E., Sanford, W. E., Connor, C. B., Sandberg, S. K. and Diez, M. 2007. Investigation of the groundwater system at Masaya Caldera, Nicaragua, using transient electromagnetics and numerical simulation. *Journal of Volcanology and Geothermal Research,* 166(3–4), 217-232.

Magaña, V., Amador, J. A. and Medina, S. 1999. The midsummer drought over Mexico and Central America. *Journal of Climate,* 12(6), 1577-1588.

Marshall, J. S., 2007. The geomorphology and physiographic provinces of Central America. *In:* Bundschuh, J. and Alvarado, G. E. eds. London, UK: Taylor and Francis, 75-122.

Martinelli, L. A., Victoria, R. L., Silveira Lobo Sternberg, L., Ribeiro, A. and Zacharias Moreira, M. 1996. Using stable isotopes to determine sources of evaporated water to the atmosphere in the Amazon basin. *Journal of Hydrology,* 183(3–4), 191-204.

McBirney, A. R. and Williams, H., 1965. *Volcanic history of Nicaragua.* University of California Press.

McClain, M. E., Boyer, E., Dent, C., Gergel, S., Grimm, N., Groffman, P., Hart, S., Harvey, J., Johnston, C. and Mayorga, E. 2003. Biogeochemical hot spots and hot moments at the interface of terrestrial and aquatic ecosystems. *Ecosystems,* 6(4), 301-312.

McClain, M. E., Chícharo, L., Fohrer, N., Gaviño Novillo, M., Windhorst, W. and Zalewski, M. 2012. Training hydrologists to be ecohydrologists and play a leading role in environmental problem solving. *Hydrol. Earth Syst. Sci.,* 16(6), 1685-1696.

McClain, M. E., Subalusky, A. L., Anderson, E. P., Dessu, S. B., Melesse, A. M., Ndomba, P. M., Mtamba, J. O. D., Tamatamah, R. A. and Mligo, C. 2013. Comparing flow regime, channel hydraulics and biological communities to infer flow-ecology relationships in the Mara River of Kenya and Tanzania. *Hydrological Sciences Journal.*

McDonnell, J. J. 1990. A Rationale for Old Water Discharge Through Macropores in a Steep, Humid Catchment. *Water Resources Research,* 26(11), 2821-2832.

McDonnell, J. J. and Woods, R. 2004. On the need for catchment classification. *Journal of Hydrology,* 299(1–2), 2-3.

Mendoza, J. A. and Barmen, G. 2006. Assessment of groundwater vulnerability in the Río Artiguas basin, Nicaragua. *Environmental Geology,* 50(4), 569-580.

Mendoza, J. A., Dahlin, T. and Barmen, G. 2006. Hydrogeological and hydrochemical features of an area polluted by heavy metals in central Nicaragua. *Hydrogeology Journal,* 14(6), 1052-1059.

Mendoza, J. A., Ulriksen, P., Picado, F. and Dahlin, T. 2008. Aquifer interactions with a polluted mountain river of Nicaragua. *Hydrological Processes,* 22(13), 2264-2273.

Miller, D. N., Bryant, J. E., Madsen, E. L. and Ghiorse, W. C. 1999. Evaluation and Optimization of DNA Extraction and Purification Procedures for Soil and Sediment Samples. *Applied and Environmental Microbiology,* 65(11), 4715-4724.

Minasny, B. and Hartemink, A. E. 2011. Predicting soil properties in the tropics. *Earth-Science Reviews,* 106(1–2), 52-62.

Mitsch, W. J. and Hernandez, M. E. 2013. Landscape and climate change threats to wetlands of North and Central America. *Aquatic Sciences,* 75(1), 133-149.

Moncrieff, J., Bentley, L. and Calderon, H. 2008. Investigating pesticide transport in the León-Chinandega aquifer, Nicaragua. *Hydrogeology Journal,* 16(1), 183-197.

Moratti, J., Moraes, J. M., Rodrigues, J., J.C., Victoria, R. L. and Martinelli, L. A. 1997. Hydrograph separation of the amazon river using 18O as an isotopic tracer. *Scientia Agricola,* 54, 167-173.

Mul, M. L., Mutiibwa, R. K., Foppen, J. W. A., Uhlenbrook, S. and Savenije, H. H. G. 2007. Identification of groundwater flow systems using geological mapping and chemical spring analysis in South Pare Mountains, Tanzania. *Physics and Chemistry of the Earth, Parts A/B/C,* 32(15–18), 1015-1022.

Mul, M. L., Mutiibwa, R. K., Uhlenbrook, S. and Savenije, H. H. G. 2008. Hydrograph separation using hydrochemical tracers in the Makanya catchment, Tanzania. *Physics and Chemistry of the Earth, Parts A/B/C,* 33(1–2), 151-156.

Muñoz-Villers, L. E. and McDonnell, J. J. 2013. Land use change effects on runoff generation in a humid tropical montane cloud forest region. *Hydrology and Earth System Sciences,* 17(9), 3543-3560.

Munyaneza, O., Wenninger, J. and Uhlenbrook, S. 2012. Identification of runoff generation processes using hydrometric and tracer methods in a meso-scale catchment in Rwanda. *Hydrology and Earth System Sciences,* 16(7), 1991-2004.

Niedzialek, J. M. and Ogden, F. L. 2012. First-order catchment mass balance during the wet season in the Panama Canal Watershed. *Journal of Hydrology,* 462–463(0), 77-86.

OEA, W., DC 1992. Honduras: proyecto de manejo de los recursos naturales renovables de la Cuenca del Embalse El Cajón: estudio de factibilidad (Honduras: renewable resources management plan for the El Cajon watershed: factibility study).

Okano, Y., Hristova, K. R., Leutenegger, C. M., Jackson, L. E., Denison, R. F., Gebreyesus, B., Lebauer, D. and Scow, K. M. 2004. Application of real-time PCR to study effects of ammonium on population size of ammonia-oxidizing bacteria in soil. *Applied and Environmental Microbiology,* 70(2), 1008-1016.

Oquist, P., 2013. *Water, energy and the Grand Interoceanic Canal in the transformation of Nicaragua.*

Orozco, P. P., Brown, S., Lantagne, S., Faustino, J. and ASDI, S. J. C., Turrialba 2008. Proceso de planificación para el manejo, gestión y cogestión de la parte alta de la subcuenca del río Viejo, Nicaragua El caso del proyecto MARENA-PIMCHAS. *Seminario Internacional" Cogestión de Cuencas Hidrográficas Experiencias y Desafíos". Turrialba (Costa Rica). 14-16 Oct 2008.*

Oxtobee, J. P. A. and Novakowski, K. S. 2002. A field investigation of groundwater/surface water interaction in a fractured bedrock environment. *Journal of Hydrology,* 269(3-4), 169-193.

Parello, F., Aiuppa, A., Calderon, H., Calvi, F., Cellura, D., Martinez, V., Militello, M., Vammen, K. and Vinti, D. 2008. Geochemical characterization of surface waters and groundwater resources in the Managua area (Nicaragua, Central America). *Applied Geochemistry,* 23(4), 914-931.

Payne, B. and Yurtsever, Y., 1974. *Environmental isotopes as a hydrogeological tool in Nicaragua.* Department of Research and Isotopes, IAEA.

Perez, M., Siria, I., Sotelo, M. and Robleto, R., 2008. *Estudio de pre-factibilidad sobre produccion y comercialización de Anadara tuberculosa y Anadara similis en los manglares del municipio de Tola, Rivas (Pre-factibility study on the production and comercialization of Anadara tuberculosa and Anadara similis in the Tola Mangroves, Rivas).* Managua: GTZ.

Pfister, L., McDonnell, J. J., Wrede, S., Hlúbiková, D., Matgen, P., Fenicia, F., Ector, L. and Hoffmann, L. 2009. The rivers are alive: on the potential for diatoms as a tracer of water source and hydrological connectivity. *Hydrological Processes,* 23(19), 2841-2845.

Picado, F., Mendoza, A., Cuadra, S., Barmen, G., Jakobsson, K. and Bengtsson, G. 2010. Ecological, groundwater, and human health risk assessment in a mining region of Nicaragua. *Risk analysis,* 30(6), 916-933.

Pinder, G. F. and Sauer, S. P. 1971. Numerical Simulation of Flood Wave Modification Due to Bank Storage Effects. *Water Resources Research,* 7(1), 63-70.

Plata, A., Gonfiantini, R. and Lopez, A. 1994. Assessment of contamination risks of Asososca Lake (Nicaragua). *IAHS Publications-Series of Proceedings and Reports-Intern Assoc Hydrological Sciences,* 222, 177-188.

Plummer, L., Sanford, W. and Glynn, P., 2013. Characterization and conceptualization of groundwater flow systems. *Isotope methods for dating old groundwater.* Vienna: IAEAA, 5-19.

Plummer, L. N., Bexfield, L., Anderholm, S., Sanford, W. and Busenberg, E. 2004. Hydrochemical tracers in the middle Rio Grande Basin, USA: 1. Conceptualization of groundwater flow. *Hydrogeology Journal,* 12(4), 359-388.

PNUD and OMM, 1972. *Proyecto Hidrometeorológico Centroamericano (Central American Hydrometeorological Project).* Costa Rica.

Polidoro, B. A., Carpenter, K. E., Collins, L., Duke, N. C., Ellison, A. M., Ellison, J. C., Farnsworth, E. J., Fernando, E. S., Kathiresan, K. and Koedam, N. E. 2010. The loss of species: mangrove extinction risk and geographic areas of global concern. *PLoS One,* 5(4), e10095.

Pool, D. J., Snedaker, S. C. and Lugo, A. E. 1977. Structure of mangrove forests in Florida, Puerto Rico, Mexico, and Costa Rica. *Biotropica,* 195-212.

Praamsma, T., Novakowski, K., Kyser, K. and Hall, K. 2009. Using stable isotopes and hydraulic head data to investigate groundwater recharge and discharge in a fractured rock aquifer. *Journal of Hydrology,* 366(1–4), 35-45.

Ptak, T., Piepenbrink, M. and Martac, E. 2004. Tracer tests for the investigation of heterogeneous porous media and stochastic modelling of flow and transport—a review of some recent developments. *Journal of Hydrology,* 294(1–3), 122-163.

Rabinowitz, D. 1978. Early growth of mangrove seedlings in Panama, and an hypothesis concerning the relationship of dispersal and zonation. *Journal of Biogeography,* 113-133.

Rawls, W., David, G., Mullen, J. and Ward, T., 1996. *Hydrology Handbook.* Second ed.

Reef, R., Feller, I. C. and Lovelock, C. E. 2010. Nutrition of mangroves. *Tree Physiology,* 30(9), 1148-1160.

Rhodes, A. L., Guswa, A. J. and Newell, S. E. 2006. Seasonal variation in the stable isotopic composition of precipitation in the tropical montane forests of Monteverde, Costa Rica. *Water Resources Research,* 42(11), W11402.

Rivera-Monroy and Twilley, R. R. 1996. The relative role of denitrification and immobilization in the fate of inorganic nitrogen in mangrove sediments (Terminos Lagoon, Mexico). *Limnology and Oceanography,* 41(2), 284-296.

Rivera-Monroy, V., Twilley, R., Boustany, R., Day, J., Veraherrera, F. and Ramirez, M. 1995. Direct denitrification of mangorve sediments in terminos Lagoon, Mexico. *Marine Ecology Progress Series,* 126, 97-109.

Roa-García, M. and Weiler, M. 2010. Integrated response and transit time distributions of watersheds by combining hydrograph separation and long-term transit time modeling. *Hydrology and Earth System Sciences,* 14(8), 1537-1549.

Roth, L. C. 1992. Hurricanes and mangrove regeneration: effects of Hurricane Joan, October 1988, on the vegetation of Isla del Venado, Bluefields, Nicaragua. *Biotropica,* 375-384.

Rubin, Y. and Hubbard, S. S., 2006. *Hydrogeophysics.* Springer.

Sabir, I. H., Haldorsen, S., Torgersen, J., Alestrom, P., Gaut, S., Colleuille, H., Pedersen, T. S., Kitterod, N.-O. and Alestrom, P. 2000. Synthetic DNA tracers: examples of their application in water related studies. *IAHS Publication(International Association of Hydrological Sciences),* (262), 159-165.

Sabir, I. H., Torgersen, J., Haldorsen, S. and Aleström, P. 1999. DNA tracers with information capacity and high detection sensitivity tested in groundwater studies. *Hydrogeology Journal,* 7(3), 264-272.

Saenger, P., 2002. *Mangrove ecology, silviculture and conservation.* Netherlands: Springer

Saha, D., Dwivedi, S. N., Roy, G. and Reddy, D. V. 2013. Isotope-based investigation on the groundwater flow and recharge mechanism in a hard-rock aquifer system: the case of Ranchi urban area, India. *Hydrogeology Journal,* 21(5), 1101-1115.

Salemi, L. F., Groppo, J. D., Trevisan, R., de Moraes, J. M., de Barros Ferraz, S. F., Villani, J. P., Duarte-Neto, P. J. and Martinelli, L. A. 2013. Land-use change in the Atlantic rainforest region: Consequences for the hydrology of small catchments. *Journal of Hydrology,* 499(0), 100-109.

Savenije, H. H. G. 1995. New definitions for moisture recycling and the relationship with land-use changes in the Sahel. *Journal of Hydrology,* 167(1–4), 57-78.

Savenije, H. H. G. 2004. The importance of interception and why we should delete the term evapotranspiration from our vocabulary. *Hydrological Processes,* 18(8), 1507-1511.

Savenije, H. H. G. 2009. HESS Opinions "The art of hydrology"*. *Hydrol. Earth Syst. Sci.,* 13(2), 157-161.

Savenije, H. H. G. 2010. HESS Opinions" Topography driven conceptual modelling (FLEX-Topo)". *Hydrology and Earth System Sciences,* 14(12), 2681-2692.

Schaap, M. G., Leij, F. J. and van Genuchten, M. T. 2001. rosetta: a computer program for estimating soil hydraulic parameters with hierarchical pedotransfer functions. *Journal of Hydrology,* 251(3–4), 163-176.

Schäfer, H. and Muyzer, G., 2001. Denaturing gradient gel electrophoresis in marine microbial ecology. *In:* John, H. P. ed. *Methods in Microbiology.* Academic Press, 425-468.

Schefe, J., Lehmann, K., Buschmann, I., Unger, T. and Funke-Kaiser, H. 2006. Quantitative real-time RT-PCR data analysis: current concepts and the novel "gene expression's C T difference" formula. *Journal of Molecular Medicine,* 84(11), 901-910.

Scheibye, K., Weisser, J., Borggaard, O. K., Larsen, M. M., Holm, P. E., Vammen, K. and Christensen, J. H. 2014. Sediment baseline study of levels and sources of polycyclic aromatic hydrocarbons and heavy metals in Lake Nicaragua. *Chemosphere,* 95(0), 556-565.

Schellekens, J., Scatena, F., Bruijnzeel, L., Van Dijk, A., Groen, M. and Van Hogezand, R. 2004. Stormflow generation in a small rainforest catchment in the Luquillo Experimental Forest, Puerto Rico. *Hydrological Processes,* 18(3), 505-530.

Schosinsky, G. and Losilla, M. 2000. Modelo analítico para determinar la infiltración con base en la lluvia mensual (Analytical model to determine infiltration based on monthly precipitation). *Revista Geológica de América Central,* 23, 43-55.

Schumacher, M., 2007. *Formulación e implementación de medidas de conservación de manglares (Design and implementation of conservation actions for mangrove forests).* Managua: GTZ.

Seo, S. B., Jin, H. X., Lee, H. Y., Ge, J., King, J. L., Lyoo, S. H., Shin, D. H. and Lee, S. D. 2013. Improvement of short tandem repeat analysis of samples highly contaminated by humic acid. *Journal of forensic and legal medicine,* 20(7), 922-928.

Sequeira Gómez, L. and Escolero Fuentes, O. 2010. The application of electrical methods in exploration for ground water resources in the River Malacatoya sub-basin, Nicaragua. *Geofísica internacional,* 49(1), 27-41.

Sharma, A. N., Luo, D. and Walter, M. T. 2012. Hydrological Tracers Using Nanobiotechnology: Proof of Concept. *Environmental science & technology,* 46(16), 8928-8936.

Sherman, R. E., Fahey, T. J. and Howarth, R. W. 1998. Soil-plant interactions in a neotropical mangrove forest: iron, phosphorus and sulfur dynamics. *Oecologia,* 115(4), 553-563.

SICA, 2010. *Estrategia regional de cambio climatico (Regional strategy for climate chage).* El Salvador: SICA.

Silliman, S. E. and Booth, D. F. 1993. Analysis of time-series measurements of sediment temperature for identification of gaining vs. losing portions of Juday Creek, Indiana. *Journal of Hydrology,* 146, 131-148.

Šimůnek, J., van Genuchten, M. T., Gribb, M. M. and Hopmans, J. W. 1998. Parameter estimation of unsaturated soil hydraulic properties from transient flow processes. *Soil and Tillage Research,* 47(1–2), 27-36.

Šimůnek, J., van Genuchten, M. T. and Sejna, M., 2012. *The HYDRUS-2D software package for simulating two-dimensional movement of water, heat, and multiple solutes in variably-saturated media, version 2.0.* Prague: PC Progress.

SINAPRED, 2005. Plan de gestion de riesgos. Departamento de Rivas. Municipio de San Juan del Sur. (Risk management plan. Rivas Department. Municipality of San Juan del Sur). [online].

Singh, R., Archfield, S. A. and Wagener, T. 2014. Identifying dominant controls on hydrologic parameter transfer from gauged to ungauged catchments – A comparative hydrology approach. *Journal of Hydrology,* 517(0), 985-996.

Sivapalan, M., 2006. Pattern, Process and Function: Elements of a Unified Theory of Hydrology at the Catchment Scale. *Encyclopedia of Hydrological Sciences.* John Wiley & Sons, Ltd.

Sivapalan, M., Savenije, H. H. G. and Blöschl, G. 2012. Socio-hydrology: A new science of people and water. *Hydrological Processes,* 26(8), 1270-1276.

Sivapalan, M., Takeuchi, K., Franks, S. W., Gupta, V. K., Karambiri, H., Lakshmi, V., Liang, X., McDonnell, J. J., Mendiondo, E. M., O'Connell, P. E., Oki, T., Pomeroy, J. W., Schertzer, D., Uhlenbrook, S. and Zehe, E. 2003. IAHS Decade on Predictions in Ungauged Basins (PUB), 2003–2012: Shaping an exciting future for the hydrological sciences. *Hydrological Sciences Journal,* 48(6), 857-880.

Sklash, M. G. and Farvolden, R. N. 1979. The role of groundwater in storm runoff. *Journal of Hydrology,* 43(1), 45-65.

Smith, A. P., Hogan, K. P. and Idol, J. R. 1992. Spatial and temporal patterns of light and canopy structure in a lowland tropical moist forest. *Biotropica,* 503-511.

Snedaker, S. C., 1993. Impact on mangroves. *In:* Maul, G. A. ed. *Climatic change in the Intra-American Seas: implication of future climate change on the ecosystems and socio-economic structure of the marine regimes of the Caribbean Sea, Gulf of Mexico, Bahamas and NE Coast of S. America.* 1st ed. London, 282-305.

119

Sophocleous, M. 2002. Interactions between groundwater and surface water: the state of the science. *Hydrogeology Journal,* 10(1), 52-67.

Sophocleous, M. 2007. The Science and Practice of Environmental Flows and the Role of Hydrogeologists. *Ground Water,* 45(4), 393-401.

Sophocleous, M. A. 1991. Stream-floodwave propagation through the Great Bend alluvial aquifer, Kansas: Field measurements and numerical simulations. *Journal of Hydrology,* 124(3–4), 207-228.

Souza Filho, P. W. M. and Paradella, W. R. 2002. Recognition of the main geobotanical features along the Bragança mangrove coast (Brazilian Amazon Region) from Landsat TM and RADARSAT-1 data. *Wetlands Ecology and Management,* 10(2), 121-130.

Spalding, M., Blasco, F. and Field, C. D., 1997. *World mangrove atlas.* UK: Earthscan.

Squillace, P. J. 1996. Observed and Simulated Movement of Bank-Storage Water. *Ground Water,* 34(1), 121-134.

Stocker, T. F., Qin, D., Plattner, G.-K., Tignor, M., Allen, S. K., Boschung, J., Nauels, A., Xia, Y., Bex, V. and Midgley, P. M., 2013. *Climate change 2013: The physical science basis.* New York.

Stubner, S. 2002. Enumeration of 16S rDNA of< i> Desulfotomaculum</i> lineage 1 in rice field soil by real-time PCR with SybrGreen™ detection. *Journal of Microbiological Methods,* 50(2), 155-164.

Stuyfzand, P. J. 1999. Patterns in groundwater chemistry resulting from groundwater flow. *Hydrogeology Journal,* 7(1), 15-27.

Suzuki, M. T., Taylor, L. T. and DeLong, E. F. 2000. Quantitative analysis of small-subunit rRNA genes in mixed microbial populations via 5 -nuclease assays. *Applied and Environmental Microbiology,* 66(11), 4605-4614.

Swain, F. 1966. Bottom sediments of lake Nicaragua and lake Managua, Western Nicaragua. *Journal of Sedimentary Research,* 36(2), 522-540.

Takai, K. and Horikoshi, K. 2000. Rapid detection and quantification of members of the archaeal community by quantitative PCR using fluorogenic probes. *Applied and Environmental Microbiology,* 66(11), 5066-5072.

Tan, X., Kong, F., Zeng, Q., Cao, H., Qian, S. and Zhang, M. 2009. Seasonal variation of Microcystis in Lake Taihu and its relationships with environmental factors. *Journal of environmental sciences (China),* 21(7), 892.

Thom, B. G. 1967. Mangrove ecology and deltaic geomorphology: Tabasco, Mexico. *The Journal of Ecology,* 55(2), 301-343.

Thornthwaite, C. W. and Mather, J. R. 1957. Instructions and tables for computing potential evapotranspiration and the water balance. *Publications in Climatology,* 10(3), 185-311.

Tóth, J. 1963. A theoretical analysis of groundwater flow in small drainage basins. *Journal of Geophysical Research,* 68(16), 4795-4812.

Tweed, S. O., Weaver, T. R. and Cartwright, I. 2005. Distinguishing groundwater flow paths in different fractured-rock aquifers using groundwater chemistry: Dandenong Ranges, southeast Australia. *Hydrogeology Journal,* 13(5-6), 771-786.

Uhlenbrook, S. 2006. Catchment hydrology—a science in which all processes are preferential. *Hydrological Processes,* 20(16), 3581-3585.

Uhlenbrook, S. 2007. Biofuel and water cycle dynamics: what are the related challenges for hydrological processes research? *Hydrological Processes,* 21(26), 3647-3650.

Uhlenbrook, S., Didszun, J. and Wenninger, J. 2008. Source areas and mixing of runoff components at the hillslope scale—a multi-technical approach. *Hydrological Sciences Journal,* 53(4), 741-753.

Uhlenbrook, S., Frey, M., Leibundgut, C. and Maloszewski, P. 2002. Hydrograph separations in a mesoscale mountainous basin at event and seasonal timescales. *Water Resources Research,* 38(6), 31-31-31-14.

Uhlenbrook, S. and Hoeg, S. 2003. Quantifying uncertainties in tracer-based hydrograph separations: a case study for two-, three- and five-component hydrograph separations in a mountainous catchment. *Hydrological Processes,* 17(2), 431-453.

Uhlenbrook, S. and Leibundgut, C. 2002. Process-oriented catchment modelling and multiple-response validation. *Hydrological Processes,* 16(2), 423-440.

UNA, 2003. *Actualización del estado del recurso suelo y capacidad de uso de la tierra del municipio de San Juan del Sur.* Managua: Universidad Nacional Agraria.

Vammen, K., Hurtado, I., Picado, F., Flores, Y., Calderón, H., Delgado, V., Flores, S., Caballero, Y., Jimenez, M. and Saenz, R., 2012. Recursos hídricos en Nicaragua: una visión estratégica (Water resources in Nicaragua: a strategic vision). *In:* Cisneros, B. and Tundisi, J. eds. *Diagnóstico del agua en las Américas.* Mexico: IANAS, 359-403.

Vidon, P. 2012. Towards a better understanding of riparian zone water table response to precipitation: surface water infiltration, hillslope contribution or pressure wave processes? *Hydrological Processes,* 26(21), 3207-3215.

Wagener, T., Sivapalan, M., Troch, P. and Woods, R. 2007. Catchment Classification and Hydrologic Similarity. *Geography Compass,* 1(4), 901-931.

Walraevens, K. and Van Camp, M., 2005. Advances in understanding natural groundwater quality controls in coastal aquifers. *In:* Araguas, L., Custodio, E. and Manzano, M. eds. *Groundwater and Saline Intrusion. Selected papers from the 18th SWIM meeting.* 1st ed. Madrid: IGGM, 449-463.

Watson, J. D., 1992. *Recombinant DNA.* New York: Scientific American Books : Distributed by W.H. Freeman.

Waylen, P. and Sadí Laporte, M. 1999. Flooding and the El Niño-Southern Oscillation phenomenon along the Pacific coast of Costa Rica. *Hydrological Processes,* 13(16), 2623-2638.

Weeda, R., 2011. *Hydrogeobiochemistry in a small tropical delta.* Master thesis. Vrije Universiteit.

Weiler, M., McDonnell, J. J., Tromp-van Meerveld, I. and Uchida, T., 2006. Subsurface Stormflow. *Encyclopedia of Hydrological Sciences.* John Wiley & Sons, Ltd.

Welch, C., Cook, P. G., Harrington, G. A. and Robinson, N. I. 2013. Propagation of solutes and pressure into aquifers following river stage rise. *Water Resources Research,* 49(9), 5246-5259.

Wels, C., Cornett, R. J. and Lazerte, B. D. 1991. Hydrograph separation: A comparison of geochemical and isotopic tracers. *Journal of Hydrology,* 122(1–4), 253-274.

Wenjie, L., Wenyao, L., Hongjian, L., Wenping, D. and Hongmei, L. 2011. Runoff generation in small catchments under a native rain forest and a rubber plantation in Xishuangbanna, southwestern China. *Water and Environment Journal,* 25(1), 138-147.

Wenninger, J., Uhlenbrook, S., Lorentz, S. and Leibundgut, C. 2008. Identification of runoff generation processes using combined hydrometric, tracer and geophysical methods in a headwater catchment in South Africa / Identification des processus de formation du débit en combinat la méthodes hydrométrique, traceur et géophysiques dans un bassin versant sud-africain. *Hydrological Sciences Journal,* 53(1), 65-80.

Wenninger, J., Uhlenbrook, S., Tilch, N. and Leibundgut, C. 2004. Experimental evidence of fast groundwater responses in a hillslope/floodplain area in the Black Forest Mountains, Germany. *Hydrological Processes,* 18(17), 3305-3322.

Westerberg, I., Guerrero, J. L., Seibert, J., Beven, K. J. and Halldin, S. 2011. Stage-discharge uncertainty derived with a non-stationary rating curve in the Choluteca River, Honduras. *Hydrological Processes,* 25(4), 603-613.

Westerberg, I., Walther, A., Guerrero, J. L., Coello, Z., Halldin, S., Xu, C. Y., Chen, D. and Lundin, L. C. 2010. Precipitation data in a mountainous catchment in Honduras: quality assessment and spatiotemporal characteristics. *Theoretical and Applied Climatology,* 101(3-4), 381-396.

Westerberg, I. K., Gong, L., Beven, K. J., Seibert, J., Semedo, A., Xu, C. Y. and Halldin, S. 2014. Regional water balance modelling using flow-duration curves with observational uncertainties. *Hydrol. Earth Syst. Sci.,* 18(8), 2993-3013.

Westhoff, M. C., Savenije, H. H. G., Luxemburg, W. M. J., Stelling, G. S., van de Giesen, N. C., Selker, J. S., Pfister, L., Uhlenbrook, S. 2007. A distributed stream temperature model using high resolution temperature observations. *Hydrol. Earth Syst. Sci. Discuss.,* 4(1), 125-149.

Wilson, I. G. 1997. Inhibition and facilitation of nucleic acid amplification. *Applied and Environmental Microbiology,* 63(10), 3741-3751.

Winter, T. C., 1998. Ground water and surface water: A single resource. Denver: US Geological Survey (Denver, CO), 79.

Winter, T. C. 1999. Relation of streams, lakes, and wetlands to groundwater flow systems. *Hydrogeology Journal,* 7(1), 28-45.

Winter, T. C., Rosenberry, D. O. and LaBaugh, J. W. 2003. Where Does the Ground Water in Small Watersheds Come From? *Ground Water,* 41(7), 989-1000.

Woessner, W. W. 2000. Stream and Fluvial Plain Ground Water Interactions: Rescaling Hydrogeologic Thought. *Ground Water,* 38(3), 423-429.

Wohl, E., Barros, A., Brunsell, N., Chappell, N. A., Coe, M., Giambelluca, T., Goldsmith, S., Harmon, R., Hendrickx, J. M. H., Juvik, J., McDonnell, J. and Ogden, F. 2012. The hydrology of the humid tropics. *Nature Clim. Change,* 2(9), 655-662.

Wondzell, S. M. and Gooseff, M. N., 2013. Geomorphic controls on hyporheic exchange across scales: watersheds to particles. *In:* Shroder, J. F. E. ed. *Treatis in Geomorphology.* San Diego: Academic Press, 203–218.

Zhang, T. and Fang, H. H. 2006. Applications of real-time polymerase chain reaction for quantification of microorganisms in environmental samples. *Applied Microbiology and Biotechnology,* 70(3), 281-289.

Zhou, Y. 2009. A critical review of groundwater budget myth, safe yield and sustainability. *Journal of Hydrology,* 370(1–4), 207-213.

Zimmermann, A., Zimmermann, B. and Elsenbeer, H. 2009. Rainfall redistribution in a tropical forest: Spatial and temporal patterns. *Water Resources Research,* 45(11), W11413.

About the Author

Ms Heyddy Calderon is from Nicaragua. She received her BSc. in Chemical Engineering in 2001 from the National University of Engineering in Nicaragua, her thesis was related to experimental determination of infiltration parameters in mine tailings. After completion of her BSc, Ms Calderon received a scholarship from the Canadian International Cooperation Agency (CIDA) to pursue master studies. She received her MSc in Hydrogeology from the University of Calgary in 2004 with a thesis on regional groundwater flow modeling of the most  important aquifer of Nicaragua. Starting in 2004, Ms Calderon joined the Nicaraguan Aquatic Resources Research Center at the National Autonomous University of Nicaragua (CIRA-UNAN), where she is now a researcher and adjoined lecturer in hydrology. From 2004 to 2010 Ms Calderon worked in many research projects in Nicaragua related to water resources quality and quantity, groundwater modeling, hydrochemistry and stable water isotope tracers in crater lakes, and water resources management. In 2009 she started her PhD under the guidance of Prof. dr. Stefan Uhlenbrook with the financial support of Nuffic. In 2010 she received a research grant from the International Foundation for Science (IFS). In 2011 she received a 3 year grant from Faculty for the Future Foundation to conduct her PhD. Ms Calderon has several publications in peer-reviewed journals, not all of them relate to the content of this thesis. She is also President of the National Committee of IUGG for Nicaragua and National Correspondent for the International Association of Hydrological Sciences (IAHS) for Nicaragua.

List of publications related to this study:

Calderon, H., Flores, Y., Corriols, M., Sequeira, L. and Uhlenbrook, S. in review. Integrating geophysical, tracer and hydrochemical data to conceptualize groundwater flow systems in a tropical coastal catchment. *Environmental Earth Sciences*.

Calderon, H. and Uhlenbrook, S. 2014a. Characterising the climatic water balance dynamics and different runoff components in a poorly gauged tropical forested catchment, Nicaragua. *Hydrological Sciences Journal*.

Calderon, H. and Uhlenbrook, S. 2014b. Investigation of seasonal river–aquifer interactions in a tropical coastal area controlled by tidal sand ridges. *Hydrology Earth System Science Discussions,* 11(8), 9759-9790.

Calderon, H. and Uhlenbrook, S. submitted. Lessons learned from catchment scale tracer test experiments using qPCR and natural DNA from total bacteria during rainfall-runoff events in a tropical environment. *Hydrological Processes*.

Calderon, H., Weeda, R. and Uhlenbrook, S. 2014. Hydrological and geomorphological controls on the water balance components of a mangrove forest during the dry season in the Pacific Coast of Nicaragua. *Wetlands,* 34(4), 685-697.

Other publications:

Calderon, H. and Bentley, L. R. 2007. A regional-scale groundwater flow model for the Leon-Chinandega aquifer, Nicaragua. *Hydrogeology Journal,* **15**(8), 1457-1472.

Castillo Hernández, E., **Calderón Palma, H.,** Delgado Quezada, V., Flores Meza, Y. and Salvatierra Suárez, T. 2006. Situación de los recursos hídricos en Nicaragua. *Boletín Geológico y Minero,* **117**(1), 127-146.

Moncrieff, J., Bentley, L. and **Calderon Palma, H**. 2008. Investigating pesticide transport in the León-Chinandega aquifer, Nicaragua. *Hydrogeology Journal,* **16**(1), 183-197.

Parello, F., Aiuppa, A., **Calderon, H**., Calvi, F., Cellura, D., Martinez, V., Militello, M., Vammen, K. and Vinti, D. 2008. Geochemical characterization of surface waters and groundwater resources in the Managua area (Nicaragua, Central America). *Applied Geochemistry,* **23**(4), 914-931.

Vammen, K., Hurtado, I., Picado, F., Flores, Y., **Calderón, H**., Delgado, V., Flores, S., Caballero, Y., Jimenez, M. and Saenz, R., 2012. Recursos hídricos en Nicaragua: una visión estratégica. *In:* Cisneros, B. and Tundisi, J. eds. *Diagnóstico del agua en las Américas.* Mexico: IANAS, 359-403.

Netherlands Research School for the
Socio-Economic and Natural Sciences of the Environment

D I P L O M A

For specialised PhD training

The Netherlands Research School for the
Socio-Economic and Natural Sciences of the Environment
(SENSE) declares that

Heyddy Calderon

born on 23 February 1979 in Jinotepe, Nicaragua

has successfully fulfilled all requirements of the
Educational Programme of SENSE.

Delft, 16 January 2015

the Chairman of the SENSE board

the SENSE Director of Education

Prof. dr. Huub Rijnaarts

Dr. Ad van Dommelen

The SENSE Research School has been accredited by the Royal Netherlands Academy of Arts and Sciences (KNAW)

K O N I N K L I J K E   N E D E R L A N D S E
A K A D E M I E   V A N   W E T E N S C H A P P E N

The SENSE Research School declares that Ms Heyddy Calderon has successfully fulfilled all requirements of the Educational PhD Programme of SENSE with a work load of 33 EC, including the following activities:

<u>SENSE PhD Courses</u>

o   Environmental Research in Context (2011)
o   Research Context Activity: Organisation of a side forum at the World Water Summit Water energy and food nexus within the construction of the grand interoceanic canal in Nicaragua and its impact on the socioeconomic transformation of the country, Budapest (2013)

<u>Other PhD and Advanced MSc Courses</u>

o   Training in qPCR (2011)
o   Risk based correction action - RBCA (2012)
o   Water resources management and climate change (2012)
o   Python programming (2013)
o   World history of water management (2014)

<u>Management and Didactic Skills Training</u>

o   Training of two intern students in qPCR techniques (2011)

<u>Oral Presentations</u>

o   *Investigation of groundwater-surface water interactions in the Ostional Catchment.* Research meeting of the Nicaraguan Aquatic Resources Research Center (CIRA-UNAN), 11 May 2010, Managua, Nicaragua
o   *Stream-aquifer interactions in fractured sedimentary rocks using a multi-technical approach.* UNESCO-IHE Annual PhD Seminar, 05 February 2010, Delft, The Netherlands
o   *Application of natural bacterial DNA as a hydrological tracer.* Research meeting of the Nicaraguan Aquatic Resources Research Center (CIRA-UNAN), 8 February 2012, Managua, Nicaragua
o   *Hydrological and geomorphological controls on a mangrove forest subsistence during dry season in the Pacific Coast of Nicaragua.* UNESCO-IHE Annual PhD Seminar, 23 September 2013, Delft, The Netherlands

SENSE Coordinator PhD Education

Dr. ing. Monique Gulickx

T - #0432 - 101024 - C154 - 240/170/8 - PB - 9781138027589 - Gloss Lamination